VERMITECHNOLOGY

VERMITECHNOLOGY

A. Mary Violet Christy
Lecturer
Department of Biotechnology
Vivekanandha College of Arts and Science for Women
Tiruchengode
Tamil Nadu

MJP
PUBLISHERS

Chennai   Trichy   Tirunelveli   NewDelhi

MJP Publishers

All rights reserved   No 44 Nallathambi Street,
Printed and bound in India   Triplicane, Chennai 600 005

MJP 055   © Publishers, 2019

Publisher : J.C. Pillai

To
My Parents

PREFACE

Vermitechnology involves the artificial rearing of earthworms and using them for the production of vermicompost, which is a nutrient-rich material that can be used as a fertilizer. In the process, earthworms degrade organic waste material into a useful product. This book contains the basic concepts of vermicomposting written in a simple and lucid manner and is organized in six chapters.

The first chapter deals with fundamentals of earthworms. The second chapter discusses the culture of earthworms, i.e., vermiculture. The third chapter describes in detail the different methods of production of vermicompost. Chapter four discusses the role of vermicompost in plant growth and the application of vermicompost to plants. The fifth chapter deals with the details of converting vermicompost into a marketable product. The sixth chapter discusses how vermicompost can be used for organic waste reduction.

I am grateful to Dr. M. Karunanithi, Chairman, Vivekanandha Educational Institutions, and Dr. Devatha, Principal, and Dr. Vivekanandhan, Bioscience Director, Vivekanandha College of Arts and Science for Women, Tiruchengode, for stimulating and encouraging me to write this book.

I would like to express my sincere gratitude to Mr. Peter, Librarian, Thanthai Hans Roever College, Perambalur, for making useful suggestions, which helped me in composing this book.

I feel a deep sense of gratitude to my husband Mr. K. Sekar for his suggestions and encouragement and to my daughter S. Harini for her love and the much-needed happiness which she has brought into my life.

I owe my gratitude to my beloved parents R. Arpudasamy and A. Susaimary, my father-in-law, A. Krishnan, mother-in-law, K. Pitchaiammal, my brothers, A. Anand Arockia Raj, A. Vinci and A. Rex Irudaya Raj, and A. Josphine Mangala Mary, for their kind cooperation, which helped me complete this book.

I am also thankful to MJP Publishers especially Mr. J.C. Pillai, Publisher, Mr. C. Sajeesh Kumar, Managing Editor and P.Parvath Radha, Project Coordinator, for taking keen interest in bringing out this book in its present form.

A. Mary Violet Christy

Contents

I

EARTHWORM

INTRODUCTION

The earthworms are a group of invertebrates belonging to the Phylum Annelida and Class Oligochaeta and represented by more than 1000 species. Earthworm is nocturnal and the movement is effected by the alternate contraction and relaxation of circular and longitudinal muscles. The soil particles swallowed with the food probably help in the grinding operation called mechanical digestion, since it facilitates the subsequent action of the digestive enzymes. Earthworm is a free organism and it is present in moist and dark places in mud. It respires aerobically and lacks specialized respiratory organs. It has a moist skin that serves this purpose. Earthworms are of great economic value to mankind because they improve the soil quality by their action.

Annelids are segmented worms, with each segment bearing the same fundamental structures as all the others, though minor differences can occur between some segments. By distributing organs among many segments, it becomes less dangerous to an annelid if one organ is damaged. Annelids usually add new segments as they grow older by simply making new copies of the body's last segment, a sort of efficient assembly-line construction. In annelids, blood circulates in a closed system of blood vessels; it does not at some point simply drain into open sinuses, as with the molluscs. This assures that the annelid's blood does not pool in some place in its body and for a time become useless, and that only oxygen-depleted blood is circulated back to have its oxygen replenished. Annelids are covered with a very thin, cellophanelike cuticle, which cuts down on moisture loss from the body. Annelids do not dry out as fast as molluscs.

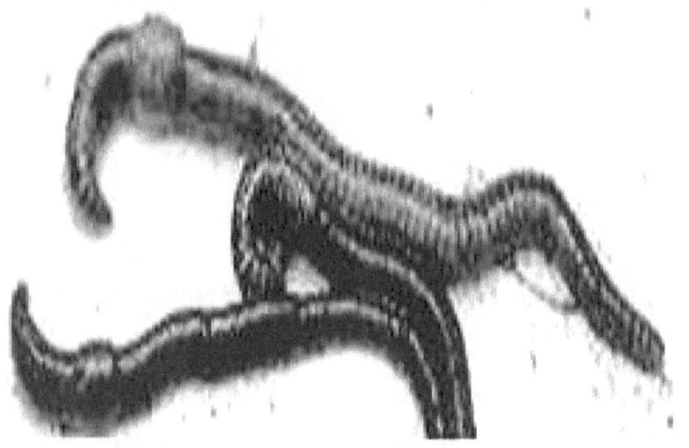

Figure 1.1   Earthworms

CATEGORIES OF EARTHWORMS

The most common earthworms in North America, Europe, and Western Asia belong to the family Lumbricidae, which has about 220 species (Figure 1.1). Earthworms ingest organic material and facilitate the redistribution of crop residues and organic matter throughout the soil profile (Timothy *et al.*, 1999). Earthworms range from a few millimetres to over 3 feet long, but most common species are a few inches in length. Only a few types are of interest to the commercial earthworm grower, and of these only two are raised on a large-scale commercial basis. In the Indian subcontinent earthworms are represented by 509 species in 67 genera under 10 families (Julka, 1993). There are more than 4400 distinct species of earthworms, each with unique physical and behavioural characteristics that distinguish them from one another. Based on the morphological nature and ecological strategies, earthworms have been classified into three groups by Bouche (1977). They are **anecic**, **endogeic**, and **epigeic**, descriptive of the area of the natural soil environment in which they are found and defined to some degree by environmental requirements and behaviours.

Anecic Species

Anecic worms feed on decaying organic matter and are responsible for cycling huge volumes of organic surface debris into humus. Represented by the common nightcrawler (*Lumbricus terrestris*), they build permanent vertical burrows that extend through the upper mineral soil layer, which can be as deep as 4–6 feet. These species coat their burrows with mucus that hardens to prevent collapse of the burrow, providing them a home to which they will always return and which they are able to reliably identify, even when surrounded by other worm burrows.

Endogeic Species

These species build extensive, largely horizontal burrow systems through all layers of the upper mineral soil. These worms rarely come to the surface, spending their lives deep in the soil where they feed on decayed organic matter and mineral soil particles. These worm species help to incorporate mineral matter into the topsoil layer as well as aerate and mix the soil through their movement and feeding habits, e.g. *Pontoscolex corethrurus.*

Epigeic Species

Represented by the common redworm (*Eisenia fetida),* these are found in the natural environment in the upper topsoil layer where they feed on decaying organic matter. Epigeic worms build no permanent burrows, preferring the loose topsoil layer rich in organic matter to the deeper mineral soil environment. Epigeic worms are the only types that are used in vermicomposting and vermiculture systems.

CHARACTERISTICS OF EARTHWORMS

Colour

Commonly earthworms are pale pinkish in colour, replaced in front by dark brown which extends backwards as a mid-dorsal stripe. The hinder end is also brown in colour. The colour of the body is due to the porphyrin present in the integument. The porphyrin protects the worm from the injurious effect of bright light.

Habitat

Earthworms are nocturnal terrestrial animals and live in burrows in the moist soil rich in dead organic matter, irrigated farmlands near the pools, ponds, rivers and gardens. They do not prefer very clayey, sandy, dry or acidic soils which are deficient in organic matter. The burrow runs almost vertically into the earth up to 45 cm deep and is built with the help of skin gland secretions. In cold weather, dried leaves and soil close the opening of the burrow. If the body surface becomes dry, the worm will die. During rainy season the burrows get flooded with water, the worms come out on the surface. During summer season the worm becomes dry. They avoid strong light. If the worms get pain by undecomposed organic material and mechanical vibrations, they come quickly to the surface.

Body Structure

The first segment is called peristomium. The next segment is prostomium. Setae are the main locomotory structures found in all body segments except first segment, last segment and clitellum (Figure 1.2). It helps the earthworm for locomotion.The muscles of the body control the movement of

Figure 1.2 Structure of earthworm

the earthworm. Mature earthworms possess a thick belt of a smooth girdle of skin around the 14th and 15th segments, called the clitellum. Depending upon the presence of the clitellum, the developmental stage of the earthworm can be divided into three stages: the mature worms called post-clitellate, the worms in the maturity-attaining stage called clitellate and immature (without clitellum) worms called pre-clitellate. The clitellum is a glandular structure that has many gland cells, which secrete the egg or cocoon. The last segment of the body is called anal segment that has a slitlike opening called anus.

Locomotion

Though earthworms have no bones, their complex system of muscles enables them to not only wiggle crazily but also to very quickly alternate between being stubby and thick, and long and slender (Figure 1.3). Earthworms possess tiny, practically invisible bristles, called setae, which usually are held inside their bodies. When the worms want to stay in their burrows, they jab their setae into the surrounding dirt, thus anchoring them in place. This comes in handy if a bird nabs a worm's head and tries to pull the worm from its burrow (Figure 1.4). The setae anchor the worm so well that it may break before coming out. If a worm in its burrow wants to move forward, first, using its complex musculature, it makes itself long. Then it anchors the front of its body by sticking its front setae into the soil. Now it pulls its rear end forward,

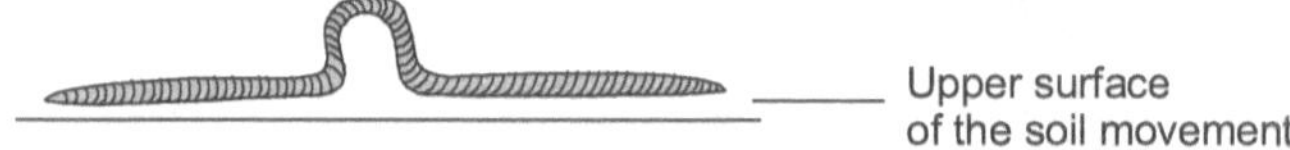

Figure 1.3   Surface movement of earthworms

Figure 1.4   Burrowing movement of earthworms

making itself short and thick. Once the rear end is in place, the front setae are withdrawn from the soil, but setae on the rear end are stuck out, anchoring the rear end. Now the front end is free to shoot forward in the burrow as the worm makes itself long and slender. Then the whole process is repeated.

Food Ingestion

The earthworm takes large amount of soil, dry leaves, grasses, algae, etc. from the earth. Earthworms prefer food material rich in nitrogen and sugar. The digestion is extracellular. The digestive fluid contains some proteolytic enzymes. Calcium is secreted by calciferous glands which probably neutralize the humic acid present in the soil. Proteins are hydrolysed into amino acids by proteolytic enzymes like trypsin and pepsin. The earthworm gut provides a suitable place for the growth of bacterial colonies and this is evidenced by the fact that earthworm castings contain significant number of bacteria, thereby increasing the number of bacteria that is present in the surrounding soil. The digested food material is passed out through the anus in the form of castings. Ejection is effected by a circular contraction which starts at the anterior end of the rectum and progresses backwards. The castings are in the form of small rounded pellets and balls.

Reproduction

Earthworms are hermaphroditic like snails and slugs, since each earthworm carries both male and female reproductive parts. First of all, not every earthworm segment bears sex organs. Counting from the front, the worm's male sex cells lie in segments 10 and 11. From here the sperms pass through sperm ducts to two male genital openings at the bottom of segment 15. On segments 9 and 10 there are two minuscule sacs called sperm receptacles, or pores, where during mating, sperms are deposited (Figure 1.5). However, this is not where eggs are produced. The egg-producing ovaries reside in segment 13, from which eggs are released through the female pores into egg sacs in segment 14. Finally, there's a rubbery, arm-band-like thin belt covering the worm's body from segments 14th and 15th, and this is called the clitellum.

Figure 1.5   Reproduction in earthworms

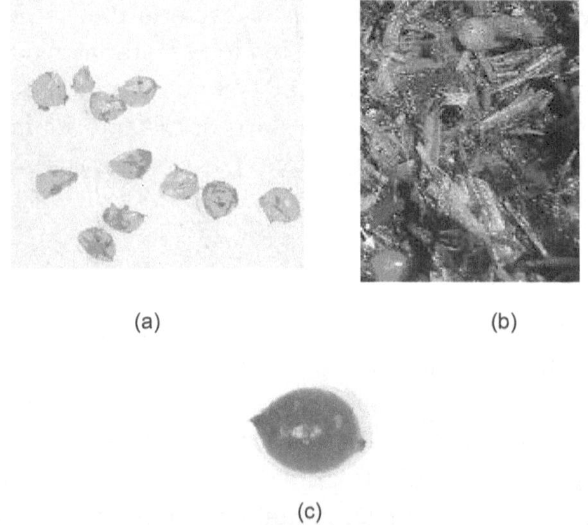

(a)   (b)

(c)

Figure 1.6   Shape   of   cocoons   of   (a)*Eisenia hortensis*, (b) *Perionyx excavatus* and (c) *Lumbricus terrestris*

Now, when two earthworms mate, they line up next to one another with their "heads" pointing in opposite directions. The clitellum of one worm lies opposite segments 9–11 of the other, containing the male parts. The worms now secrete large amounts of mucus, until each is enclosed in a slime tube extending from segment 9 to the rear end of the clitellum on segment 37. Now sperms are ejaculated from segment 15 and carried backward in tubes formed by grooves in the body touching the slime tube and the sperms pass to the sperm receptacles on segments 9 and 10 of the other worm. Then the earthworms go their different ways and mating is over. However, that is not to say that eggs have been fertilized. Each worm's sperm receptacles, receive the sperms from the other worm. Finally the cocoons are formed. The shape of the cocoons may vary depending upon the species (Figure 1.6) for instance *Eisenia hortensis*, *Perionyx excavatus* and *Lumbricus terrestris* have different shapes of cocoons.

The next step in producing young earthworms (hatchlings) takes place when the earthworm is by itself. Its clitellum secretes a second mucous ring that slides forward over the worm's body. When the ring passes the openings in segment 14, several ripe eggs leave the body and stick to the ring. The ring keeps moving forward until it passes the sperm receptacles in segments 9 and 10, and here the sperm comes in contact with eggs. Finally, within the mucous ring, the eggs are fertilized. Now the ring containing the fertilized eggs slips off the worm's head, seals at both ends, and becomes a sort of cocoon, which is left lying in the soil. Ultimately the eggs hatch and tiny worms escape from the mucous ring, into the soil.

THE BIOLOGY OF THE EARTHWORM

One of the most obvious features of the earthworm's body is its segmentation, which is not just an external feature but also occurs through almost all its internal structure. The anterior section of the earthworm, or the head, is made up of the prostomium, which is a liplike extension situated in front of the mouth, which it uses to force its way into the soil. Each of the segments in its body has hairy structures known as setae, which can be extended as well as retracted, and are used for locomotion. The absence of other locomotive structures, apart from setae, enables the earthworm to burrow efficiently into the soil. Also, there are glands in its skin which secrete mucus which help in keeping the earthworm lubricated, which again helps the earthworm to burrow easily through the soil and also aids in stabilizing casts and burrows. The earthworm has no eyes; instead, it has cells that are sensitive to light on its outer skin. These cells enable the earthworm to detect light as well as the changes in the intensity of light. The cells of the skin are sensitive to chemicals and to touch as well. The earthworm does not have any lungs either, and it breathes by oxygen being diffused into its body through the skin. The highly permeable skin of the earthworm makes it susceptible to drying out easily.

The digestive tract of the earthworm is one of its unique features, being highly adapted according to its activities of burrowing and feeding. The earthworm ingests soil, along with decomposing organic matter in it, which are mixed by powerful muscles and passed through its digestive tract. The digestive tract releases digestive fluids that contain enzymes which are mixed with the soil that has been ingested. The digestive fluids help in releasing sugars, amino acids and smaller sized organic molecules. These molecules are then absorbed via the membrane of the intestines, which are used for energy and the synthesis of new cells.

Before going into the details of vermicompost practices, one must have a good knowledge of the biology and the life cycle of suitable earthworms. Among the vast community of earthworms, only very few species can be selected or used for vermicompost. So, in this section we will discuss about the various species native to various countries suitable for vermicompost. These include *Lumbricus terrestris* (Indian nightcrawler). *Eisenia fetida* (Chinese worm), *Eudrilus eugenia* (African nightcrawler), *Amynthas gracilus* (Georgia jumper) *Perionyx excavatus* (Indian blueworm) and *Eisenia hortensis* (European nightcrawler) were described below.

Lumbricus terrestris (Indian Nightcrawler)

The Indian nightcrawler, *Lumbricus terrestris*, is used for descriptive purposes. The earthworm, while primitive, it has well-developed nervous, circulatory, digestive, excretory, muscular and reproductive systems (Figure 1.7). The most noticeable external feature is the ringing or segmentation of the body. The nightcrawler has about 150 segments. The first section of the earthworm, the anterior end or head, consists of the mouth and the prostomium, a lobe which serves as a covering for the mouth and as a wedge to force open cracks in the soil into which the earthworm may crawl. Small hairlike structures, called setae (bristles), are located on each segment. These can be extended or retracted and their principal function is for movement. The lack of protruding structures other than setae facilitates efficient burrowing; in addition, various skin glands secrete a lubricating mucus which aids movement through the earth and helps to stabilize burrows and casts.

The earthworm's digestive tract is highly adapted to its burrowing and feeding activities. The worm swallows soil (including decomposing organic residues in the soil) or residues and plant litter on the soil surface. Strong muscles mix the swallowed material and pass it through the digestive tract as digestive fluids containing enzymes are secreted and mixed with the materials. The digestive fluids release amino acids, sugars and other smaller organic molecules from the organic residues.

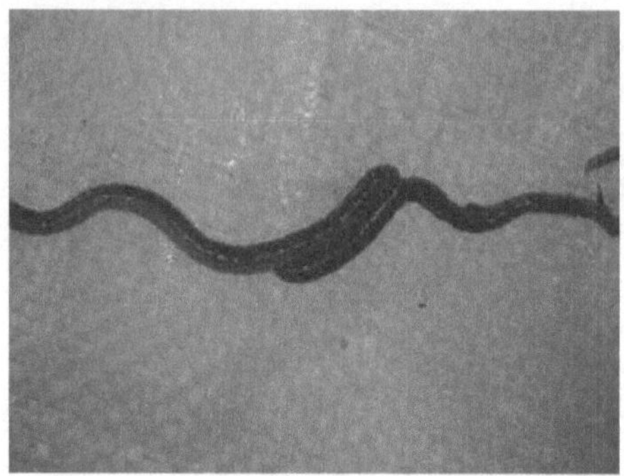

Figure 1.7 *Lumbricus terrestris*

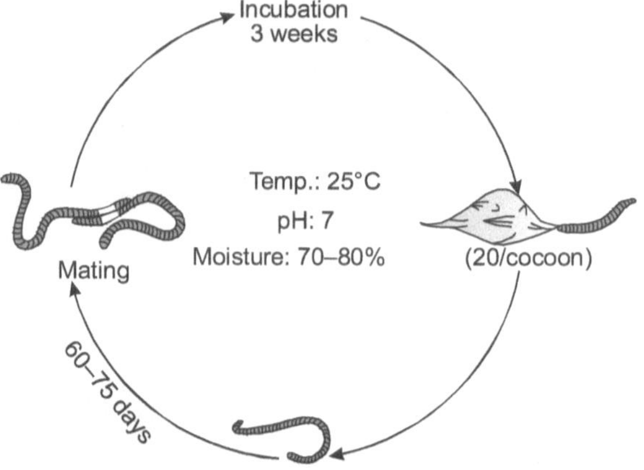

Figure 1.8 Life cycle of *Lumbricus terrestris*

The simpler molecules are absorbed through intestinal membranes and are utilized for energy and cell synthesis. Earthworms lack specialized breathing devices. Respiratory exchange occurs through the body surface. Earthworms are usually not self-mating although they are hermaphroditic (each individual possesses both male and female reproductive organs).

A mutual exchange of sperm occurs between two worms during mating. Mature sperm and egg cells and nutritive fluid are deposited in cocoons produced by the clitellum, a conspicuous, girdlelike structure near the anterior end of the body. The ova (eggs) are fertilized by the sperm cells within the cocoon, which then slips off the worm and is deposited in or on the soil. The eggs hatch after about 3 weeks, each cocoon producing from two to twenty hatchlings with an average of four. The worm cocoon is an incredibly tough structure, designed to protect the young inside from environmental extremes and even ingestion by other animals. Cocoons can be frozen, submerged in water for extended periods of time, dried and exposed to temperatures far in excess of what can be tolerated by adult worms without damage to the young worms inside (Figure 1.8). Earthworm cocoons are easy to spot in the worm bed. They are roughly the size of a large grape seed and similarly shaped, with one end rounded and the other drawn out to a point. When first dropped from the body of the parent the cocoon is a creamy, pearlescent yellow, darkening to a cola brown as the young worms within mature and prepare to emerge.

Eisenia fetida and *Eisenia andreii* (Common name, Redworm)

There are two worm species, *Eisenia fetida* and *Eisenia andreii,* called redworm. *E. andreii* closely resembles *E. fetida* in behaviour, environmental requirements, growth rate, reproduction and appearance (Figure 1.9). The only way to distinguish the difference between these two is through molecular scanning. There is no external difference between the two species. *Eisenia fetida/Eisenia andreii* are the worm species identified as the most useful in vermicomposting systems and are the easiest to grow in high-density culture because they tolerate the widest range of environmental conditions and fluctuations, and handling and disruption to their environment of all species identified for this purpose.

Figure 1.9   *Eisenia fetida*

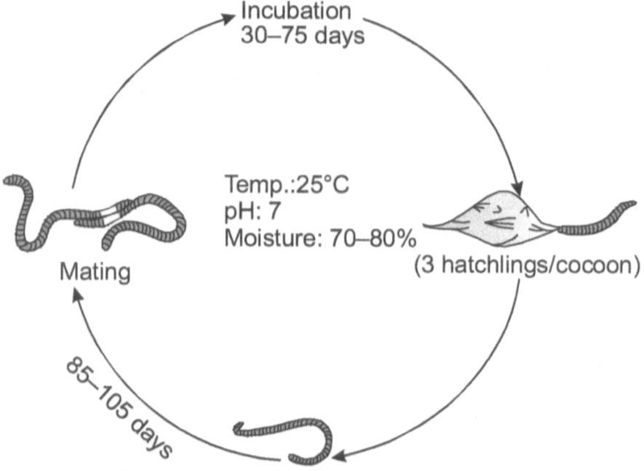

Figure 1.10   Life cycle of *Eisenia fetida*

While this worm species is considered the premier worm for most applications, it is a small worm, not always suited for use as bait. The maturation of the worms is approximately 85–150 days under ideal conditions. The worm requires an optimum temperature of 70–80°F. It can produce 10 hatchlings per worm per week under ideal conditions (Figure 1.10). Each cocoon produces an average of three hatchlings in 30–75 days. This worm is most suitable for commercial vermicomposting process.

Eudrilus eugenia
(Common name, African Nightcrawler)

This worm is a semi-tropical species, it cannot easily tolerate cool temperatures and is usually grown indoors or under temperature-controlled conditions in most areas of North America. It is well adapted to the climatic conditions of India. So, in India, when large quantities of vermicompost is to be produced *E. eugenia* species are used (Figure 1.11). *E. eugenia* is a large epigeic species and is used in the process of vermicomposting.

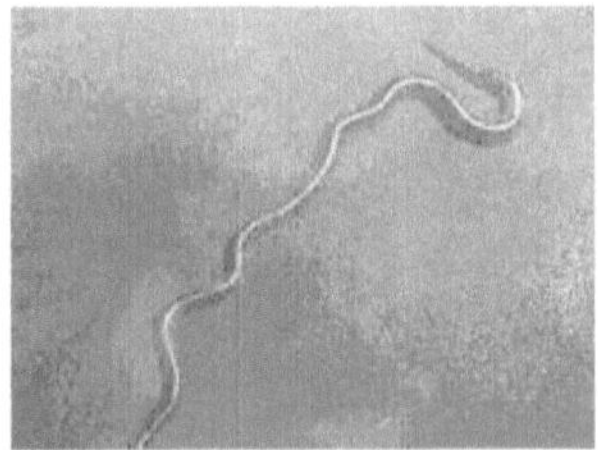

Figure 1.11　*Eudrilus eugenia*

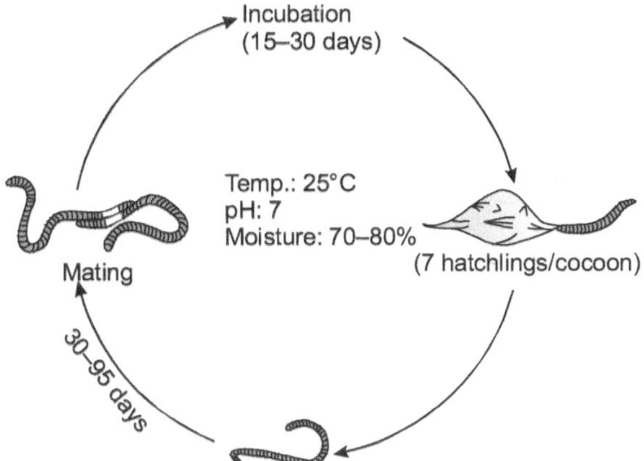

Figure 1.12　Life cycle of *Eudrilus eugenia*

The maturation of the worms is approximately 30–95 days under ideal conditions. The worm requires an optimum temperature of 70–80°F. It can produce 7 hatchlings per worm per week under ideal conditions, each cocoon produces an average of two hatchlings in 15–30 days (Figure 1.12).

Amynthas gracilus (Common name, Alabama or Georgia jumper)

A. gracilus (Figure 1.13) is another large worm species well suited for use as bait. It is also a tropical species with a poor tolerance for cold temperatures.

Figure 1.13 *Amynthas gracilus*

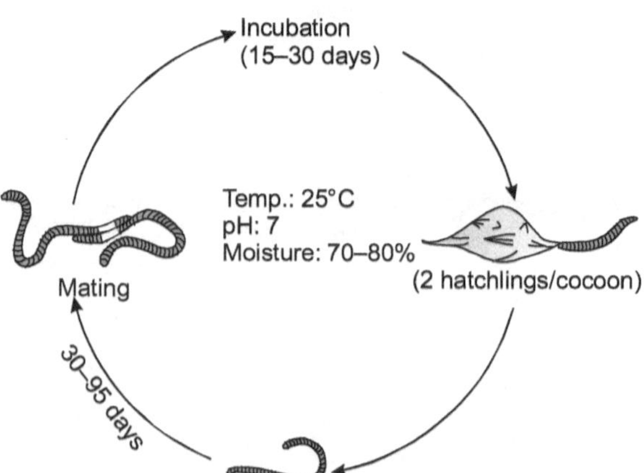

Figure 1.14 Life cycle of *Amynthas gracilus*

This worm tolerates handling and appropriate temperatures can be maintained. *A. gracilus* is disruption to the worm bed as well as does *E. fetida* and is generally considered an easy worm to culture provided used in a few vermicomposting systems in Malaysia and the Philippines. The maturation of the worms is approximately 30–95 days under ideal conditions (Figure 1.14). The worm requires an optimum temperature of 70–80°F. It can produce 7 hatchlings per worm per week under ideal conditions. Each cocoon produces an average of two hatchlings in 15–30 days.

Perionyx excavatus (Common name, Indian blueworm)

Perionyx excavatus is a beautiful worm with an iridescent blue or violet sheen to its skin clearly visible under bright light (Figure 1.15). It is an Indian earthworm which is very suitable for vermiculture and vermicomposting process. It is a very small worm, poorly suited as fishing bait, but has an impressive growth and reproductive rate far in excess of the other species grown in indoor culture. This tropical worm with a very poor tolerance of low temperatures and fluctuations in the environment.

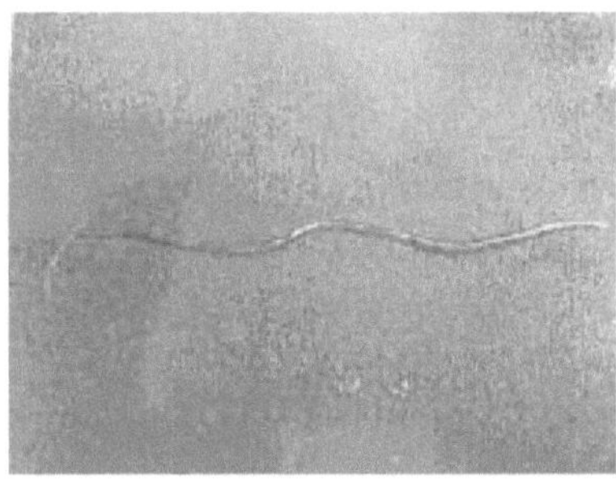

Figure 1.15 *Perionyx excavatus*

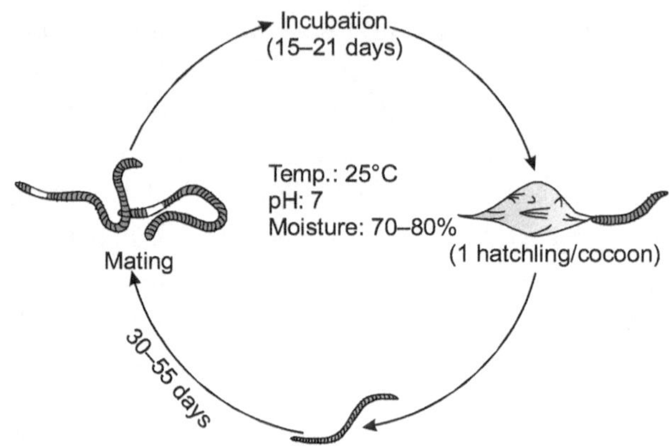

Figure 1.16 Life cycle of *Perionyx excavatus*

P. excavatus is often referred to as "the traveller" for its tendency often to leave the indoor culture. The maturation period of the worms is approximately 30–55 days under ideal conditions. The worm requires an optimum temperature of 70–80°F. It can produce 19 hatchlings per worm per week under ideal conditions (Figure 1.16). Each cocoon produces an average of one hatchling in 15–21 days.

Eisenia hortensis (European Nightcrawler)

The European *E. hortensis* is a large worm species well suited for use as a bait worm and less suited for vermicomposting process (Figure 1.17). Its ideal temperature is 20°C and it requires higher moisture level during vermicomposting, but the species tolerates handling and disruption to its environment and environmental fluctuations very well. Because this worm has a very low reproductive and growth rate, it is considered the least desirable species for the vermicomposting process. It is used in a few vermiprocessing processes in Europe for the remediation of very wet organic materials. The maturation of the worms is approximately 55–85 days under ideal conditions. The worm requires an optimum temperature of 55°–65°F. It can produce 2 hatchlings per worm per week under ideal conditions. Each cocoon produces an average of two hatchlings in 40–125 days (Figure 1.18).

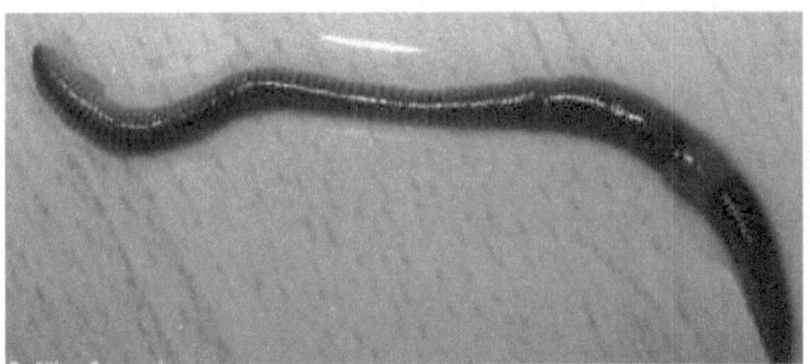

Figure 1.17   *Eisenia hortensis*

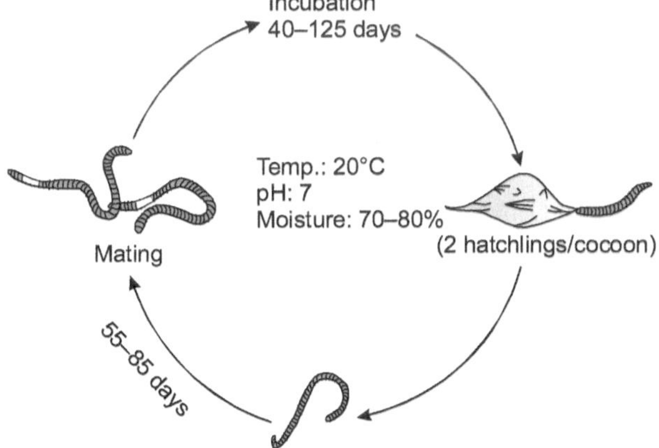

Figure 1.18   Life cycle of *Eisenia hortensis*

ADAPTATIONS OF EARTHWORMS

* The role of earthworms in the breakdown of organic matter and in the soil turnover process was highlighted by Darwin (1881).

* Earthworms are oxygen-breathing animals that absorb oxygen directly through their skin. Oxygen is dissolved into the mucus coating the worm's skin and the dissolved oxygen passes through the skin and through the walls of the capillaries of the skin where it is picked up by haemoglobin in the worm blood and carried throughout the body.

* Specific temperature requirements and tolerances vary from species to species, though the ideal range for most epigeic worm species is roughly 60–80°F. The worm's ability to tolerate temperatures outside of ideal is highly dependent on the level of moisture in the system, either hot or dry conditions.

* Breakdown of organic matter it goes through a series of naturally occurring changes in pH. Because earthworms thrive in environments rich in decaying organic matter they are adapted to tolerate these pH fluctuations with little or no change in their activity levels. In nature, worms are found in environments with a pH range from 4 to 9, with processing and reproductive rates being no different at an acidic 4 than they are at an alkaline 9. In fact, earthworms actually prefer an environment with a pH of 5 to 5.5, contrary to the popular belief that they prefer a neutral pH.

* With a pH tolerance this wide it is highly unusual for pH to be a limiting factor in worm culture. Further, the radical and artificial adjustment of the pH through the addition of buffering agents like lime can actually have a detrimental effect on the culture.

* All earthworms are photophobic to some degree, meaning they react negatively to bright light. The severity of the reaction depends on the species of worm, how bright the light and the level of light to which the worm is accustomed.

For example, earthworms accustomed to some light exposure will react less negatively to sudden bright light than will worms accustomed to complete darkness.

* Earthworms sense light through photoreceptive organs along their back and on the prostomium (sensitive lobe of tissue overhanging the mouth that the worm uses to probe and sense its environment).

* While earthworm taxonomists have identified thousands of individual worm species, only six have been identified as useful in vermicomposting practices. These species were evaluated based on their ability to tolerate a wide range of environmental conditions and fluctuations, handling and disruption to the worm bed and for their growth and breeding rate.

* Earthworm species with a short generation time, meaning a relatively short lifespan and rapid growth and reproductive rate, have been identified as most effective due in large part to the high concentration of juvenile worms present in their populations. Juvenile worms are voracious consumers, keeping the processing rate of the culture high and ensuring an ongoing succession of young worms.

* The worm's streamlined body with no conspicuous appendages is an adaptation to living in narrow burrows underground. Earthworms do not have eyes, but they do have light-sensitive cells scattered in their outer skin. Although these cells do not enable earthworms to see images, they do give their skin the capacity to detect light and changes in light intensity. The worm's skin cells are also sensitive to touch and chemicals. Earthworms have simple brains which specialize in directing body movement in response to light, and not much else.

* Earthworms have considerable powers to regenerate. If segments are removed at the posterior end a tail will be regenerated, but if the head is removed they do not regenerate the head.

IMPORTANCE OF EARTHWORMS

Earthworms have beneficial physical, chemical and biological effects on soil and can increase plant growth and crop yield. Organic wastes can be applied to soil after being processed by earthworms to obtain a mature and stabilized compost. One important thing that earthworms do is to plough the soil by tunnelling through it. Their tunnels provide the soil with passageways through which air and water can circulate, and this is important because soil microorganisms and plant roots need air and water. Without some kind of plough, soil becomes compacted, air and water cannot circulate in it, and plant roots cannot penetrate it.

Enhancement of the Soil

The improvement of soil needs decomposition and also needs two sources: the first is the mineral content of the material and the second is organic material originating from the dead parts which fall on to the surface of the soil. The decomposition process incorporates physical and chemical processes. Physical and chemical decomposition enhanced by the activity of earthworms involve the degradation of complex molecules into simpler ones.

Changes in Soil pH

Earthworms are capable of making the soil neutral or slightly alkaline in pH, which enhances the growth of bacteria and fungi, which in turn completes the degradation of organic compounds.

Agent of Decomposition

When organic material enters the gut, it initiates decomposition. The grinding action of this part of the gut may be enhanced by the ingestion of silica granules. These silica granules are commonly present in grassland plant soil systems where the earthworms are in high densities rather than in acid soils.

Humus Formation

The process is often characterized by the selective breakdown of cellulose. Lignin fibres are present in raw humus but are degraded to polyphenols in well decomposed humus. The presence of cellulose in the intestine of at least some earthworms suggests that these earthworms may play an active role in humus formation.

Improvement of Soil Quality

Earthworms swallow the earth below the surface thereby increasing the fertility of the soil. Their burrows permit the penetration of air and moisture in the porous soil, improving drainage and effecting easier downward growth of the roots. The excretory wastes and other secretions of the worms also enrich the soil by adding nitrogenous matters that form important plant food. Earthworms bring about optimum conditions for plant growth by reducing both acidity and alkalinity of the soil.

EARTHWORMS AND MICROORGANISMS

The primary decomposers of organic matter are microorganisms. Microbial activity in the earthworm's gut, cast and soil is very essential for the breakdown and release of nutrients in available form to plant from the organic wastes.

The microorganisms and earthworms act symbiotically to accelerate and enhance the decomposition of organic matter (Figure 1.19). The earthworms derive their nutrients from the organic matter that is being decomposed and also from the proliferating microorganisms. The earthworms are not directly attracted by decaying organic matter, but are actually attracted by microbes in them which act as primary consumers of the organic matter. So, earthworms can be aptly called as microbivores. Fungi and bacteria are the protein-rich food sources for earthworms (Day, 1950). Some fungi are preferentially ingested while others are rejected (Dash *et al.*, 1986). Although the bulk of food ingested by earthworms is dead plant tissues, they also feed upon living microorganisms, mesofauna and macrofauna and their dead tissue (Senapati and Dash, 1984).

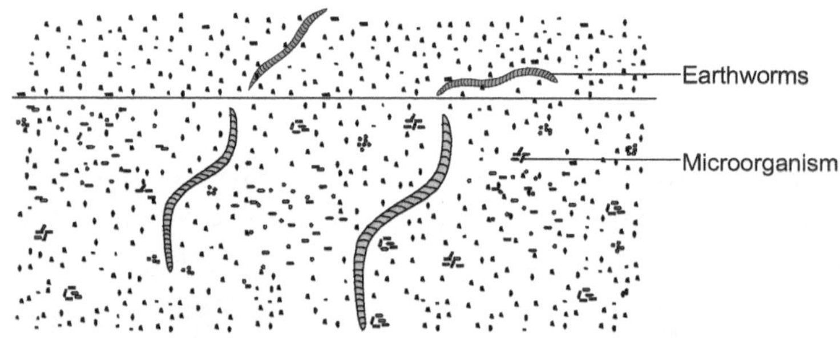

Figure 1.19   Earthworm and microorganisms in the soil

PESTS AND DISEASES OF EARTHWORMS

Compost worms are not subject to diseases caused by microorganisms, but they are subject to predation by certain animals and insects (red mites are the worst) and to a disease known as "sour crop" caused by environmental conditions.

(a) *Rat*   Earthworms are affected by rats that get access to the worm bed, and a lot of worms can be lost very quickly. This is usually a problem only when using windrows or other open-air systems in fields. It can be prevented by putting some form of barrier, such as wire mesh or paving, under the windrow.

(b) *Birds*   They are not usually a major problem, but if they discover the beds they will become regular visitors to the worm bed. Putting a windrow cover of some type over the material will eliminate this problem. These covers are also useful for retaining moisture and preventing too much leaching during rainfall.

(c) *Centipedes*   These insects eat compost worms and their cocoons. Fortunately, they do not seem to multiply to a great extent within worm beds or windrows, so damage is usually light. If they do become a problem, one method suggested for reducing their numbers is to heavily wet (but not quite flood) the worm beds. The water forces centipedes and other insect pests (but not the worms) to the surface, where they can be destroyed.

(d) *Ants*   These insects are more of a problem because they consume the feed meant for the worms (Myers, 1969). Ants are particularly attracted to sugar, so avoiding sweet feeds in the worm beds reduces this problem to a minor one.

(e) *Mites*   There are a number of different types of mites that appear in vermiculture and vermicomposting practices, but only one type is a serious problem: red mites. White and brown mites compete with worms for food and can thus have some economic impact, but red mites are parasitic on earthworms. They suck blood or body fluid from worms and they can also suck fluid from cocoons.

ECONOMIC IMPORTANCE OF EARTHWORMS

(a) *Laboratory use* The earthworms are easily obtained and are of convenient size for dissections. They are frequently used for investigations in general anatomy and comparative physiology.

(b) *Food for other animals* Earthworms are used as bait by fishing. They are also used as food in many countries. They are also eagerly hunted as food for some birds, chickens, frogs, lizards, small snakes, centipedes, etc. Earthworms are not only hunted by other animals but also by humans, because in many countries, biscuits, soups, and tablets are prepared using earthworms.

(c) *Research and development* Earthworms can rapidly multiply. They have short life cycle. They can grow with the help of cheap organic substances. They are harmless and are easy to handle and available. Earthworms are easily available, throughout the year, their cost is very low and there is no need of maintenance equipment. So, they are suitable for research and development activities.

MEDICINAL IMPORTANCE OF EARTHWORMS

(a) *Medicine for asthma* *Pheretima elongata* (Family Megascolecidae, i.e., earthworm) has been documented as a potent agent for the treatment of cough and breathing difficulty in traditional Chinese medicine for nearly 2000 years. The use of *Pheretima* (usually dried), has shown some encouraging results in combating asthmatic symptoms. The study was conducted by traditional Chinese medicine researchers in Shanghai. In the study, the water extract of *Pheretima* was separated into three fractions—the ethanolic precipitate, the alkaline fraction and the acidic fraction.

These three preparations showed better results in treating asthma.

(b) *Traditional Chinese medicine (TCM)*   According to TCM, *Pheretima* is known for its anti-asthmatic effects and diuretic effects. It is used commonly in TCM treatments. Skin of earthworms dried under the sun is used for TCM preparations. One dried earthworm *(Pheretima)* is about 15–20 cm in length and 1–1.5 cm in width. *Pheretima* is brownish in colour and smells a little fishy. Ingredients extracted from *Pheretima* include, but are not limited to, xanthine (which helps relax the airway muscles) and succinic acid (which helps calm down the respiratory system); as well as lumbritin which serves as a blood thinner. *Pheretima* is suitable for the preparation of TCM which is available in a Guangdong and Guangxi.

(c) *Medicine for antihyperlipaemic, antidiabetic and antihypertensive*   Conventional methods for producing the dried products of dried powder of earthworms are roughly divided into the following three types.

i. In the first method, the body of each earthworm is cut open, freed of its contents (i.e., the guts and soil), and then dried in the sun, in the shade or at low temperature (usually 50°C or below).

ii. In the second method, earthworms are killed by placing them in ash from plants or charcoal, freed of any soil, and then dried in the sun, in the shade or at low temperature (usually 50°C or below) to obtain dried earthworms having soil therein.

iii. In the third method, earthworms are freed of the soil present in their body and then dried by placing them in ash from plants or charcoal.

Table 1.1   The composition of dried earthworm powder products

Content	%
Moisture	10.4%
Crude proteins	53.8%
Crude lipids	12.2%
Ash	5.3%
Crude fibres	0.1%
Ca	0.384 g %
Mg	0.194 g %
K	0.613 g %
Na	0.382 g %
P	0.504 g %
Fe	0.061 g %
Cu	1.72 mg %
Zn	5.63 mg %
Mn	1.54 mg %

These methods provide a process for the production of dried earthworm powder, which a pharmaceutically acceptable and highly safe and exhibits excellent antihyperlipaemic, antidiabetic or hypoglycaemic, antihypertensive and/or antihypotensive activities without producing any side effects and can be obtained in high yield. It also provides a method for the treatment of hyperlipaemia, diabetes, hypertension or hypotension in mammals and, in particular, human beings. The administration of dried earthworm powder directly prevents the disease without producing any side effects. Although the commercial drugs available in the market for treating diabetes, hyperlipaemia and blood pressure, these drugs must be used for prolonged periods, and in addition to having excellent efficacy they also produce side effects. But dried earthworm powder is highly safe, causing no ill effects on any system of the body and/or producing no side effects. This powder can be stored in air-tight containers for 4 years. Table 1.1 shows the composition of dried earthworm powder products.

REVIEW QUESTIONS

1. Describe the characteristic features of earthworms?
2. Give an account of the biology of the earthworm?
3. List out the importance of earthworms?
4. Give an account of the pests and diseases of earthworms?
5. Discuss the economic importance of earthworms.

2

VERMICULTURE

INTRODUCTION

Vermiculture, using composting earthworms to reduce animal and plant wastes, has recently emerged as a rural industry worldwide with the expectation of a substantial market for nutrients and microbial rich vermicompost as an organic manure/fertilizer. The earthworms are used to degrade organic waste, indicate environmental pollution, detoxicate polluted soil, turn over the soil, prepare protein-rich animal feed, produce medicine, improve soil physico-chemical properties, restore soil fertility, induce plant growth, increase plant productivity, reduce pathogenic microbes, etc. The worms feed upon organic matter which undergoes physical, chemical and biological changes with the help of microorganisms.

VERMICULTURE PROCESS

Vermiculture is the culture of earthworms. The goal is to continually increase the number of worms in order to obtain a sustainable harvest. The worms are either used to expand a vermicomposting operation or sold to customers who use them for the same or other purposes. Vermiculture means artificial rearing or cultivation of worms (earthworms) and the technology is the scientific process of using them for the betterment of human beings. Vermiculture is the process of using worms to decompose organic food waste, turning the waste into a nutrient-rich material capable of supplying necessary nutrients to help sustain plant growth. This method is simple, effective, convenient, and noiseless. It saves water, energy, landfills, and helps rebuild the soil. Vermiculture composting is nature's way of completing the recycling loop. Adding compost to soil aids in erosion control, promotes soil fertility, and stimulates healthy root development in plants.

Earthworm's Food

All types of degradable organic wastes can be degraded with the help of earthworms. But it needs favourable food like cow dung and pressmud. So, any one of these organic wastes are mixed with different types of plant and animal wastes during vermiculture. The characteristic features of these wastes are given below:

(a) *Cow dung* Urine-free cow dung (sun-dried and powdered) is used as stock substrate to maintain the worms. The moisture content of the substrate is maintained at 60–75% and temperature at $27 \pm 2°C$. Organic manures serve as cheaper and effective substitutes than the chemical fertilizers. At present, the availability of traditional organic manures is limited due to various reasons. So, utilization of available organic wastes for the production of organic manures using earthworms and recycling them in agriculture in the form of vermiculture is gaining importance in recent years (Kale, 1994). The nutrient status of farmyard manure in given in Table 2.1.

Table 2.1 Nutrient profile of vermicompost and farmyard manure

Nutrient	Vermicompost	Farmyard manure
N(%)	1.6	0.5
P (%)	0.7	0.2
K (%)	0.8	0.5
Ca(%)	0.5	0.9
Mg (%)	0.2	0.2
Fe (ppm)	175.0	146.5
Mn (ppm)	96.5	69.0
Zn (ppm)	24.5	14.5
Cu (ppm)	5.0	2.8
C : N ratio	15.5	31.3

Source: Punjab State Council for Science and Technology, Chandigarh

These values are subject to variation depending upon the type of organic waste.

(b) *Pressmud* Pressmud or filtercake is one of the important by-products of sugar industries like bagasse and molasses. The production of pressmud is normally 4–5% of cane weight, 35 kg of pressmud per ton of cane processed. In Tamil Nadu 100 tons of pressmud is produced from the sugar factories per day (Solayappan *et al.*, 1997). Physically, pressmud is a soft, spongy, light-weight, amorphous, dark brown to black material containing sugar, fibre, cane wax, organic compounds, inorganic salts and soil particles. At present, pressmud is used as a source of plant nutrients, animal feed, wax, protein and carbon, carrier for legumes inoculants, feed stock material for biogas production, for making brick after burning and as a medium for raising sugar cane seedlings. Pressmud as such has a high nutrient value to be used as manure (Owen, 1954). The nutritional content of the pressmud is organic matter (53.2%), nitrogen (2.8%), phosphorus (2.4%), potassium (1.3%), calcium (2.5%), magnesium (0.6%), sulphur (0.64%), sodium (0.21%), crude protein (17.6%), zinc (62.20 mg kg^{-1}), iron (177.6 mg kg^{-1}), manganese (23.20 mg kg^{-1}), copper (40.0 mg kg^{-1}) and sugars (10%).

Utilization of Potential Organic Resources for Vermiculture

All types of organic wastes can be converted into vermicompost but meat and dairy products were not suitable for vermiculture. Redworms, *Eisenia fetida,* are especially the best type of worms for eating food waste. These worms are surface worms and stay in the top 18 inches of the soil. It is usually best to keep them in a closed container. Every three months, the worms should be harvested and separated from the castings. Worms are easy to care for but they require food, moisture, oxygen, and a dark place to live.

Some organic wastes which are suitable for vermiculture are:

1.   *Animal wastes/by-products*

   * Cattle dung
   * Buffalo dung and urine
   * Goat and sheep droppings
   * Pig slurry
   * Waste meat
   * Leather wastes

2.   *Crop residues/by-products*

   * Rice straw
   * Cereal residues
   * Pulse residues

* Bagasse
* Pressmud
* Rice husk
* Tobacco wastes
* Cotton dust
* Tea waste

3. *Fruits and vegetable wastes/residues*

4. *Forest residues/by-products*

* Forest litter
* Sawdust
* Non-edible oilcakes

5. *Fish and marine wastes/residues*

* Prawn shell and head
* Fish and frog waste

6. *Human habitation wastes*

* City refuse
* Night soil
* Human urine

7. *Aquatic biomass/other wastes*

* Water hyacinth

MONOCULTURE AND POLYCULTURE

The benefits of vermicompost mainly depends upon mono- and polyculture. The culture does not need any special type of wastes. Polyculture method was designed by using epigeic and anecic species of earthworms, while in monoculture method any one of the species utilized. Comparatively the end product, i.e., vermicompost, obtained from polyculture method shows a better concentration of nutrients and lower metal levels than monoculture method. Polyculture method showed greater dehydrogenase activity and microbial biomass, possibly related to the better microclimatic conditions in it. This method accelerates the mineralization process and also enhances the microbial activity.

Hatchlings of *Lumbricus terrestris, Allolobophora chlorotica, Aporrectodea longa* and *Lumbricus rubellus* were cultured in a pot at 15°C in darkness. Each hatchlings survive with their respective adult worms. The experiments were continued up to 6 months. The adults were analysed with normal growth parameters like percentage of survival, growth rate, maturation, cocoon production, hatchling production, etc. During the experiment 90% earthworms survived. The intensity of interactions were directly related to the degree of niche overlap between pairing. The growth of anecic and endogeic species were significantly reduced by the presence of *Lumbricus rubellus.* Growth and cocoon production of *Allolobophora chlorotica* was enhanced by the presence of *Aporrectodea longa.* During polyculture all species of earthworms survived and some species were present dominantly than the others. Care should be taken in the choice of earthworms while introducing earthworms for field study.

REVIEW QUESTIONS

1.　What is vermiculture?
2.　What are the sources for earthworm's food?

3

VERMICOMPOST AND VERMIWASH

INTRODUCTION

The process of converting wastes of plant and animal origin, by various processes such as decay, decomposition, fermentation, etc., is known as "composting". It is understood broadly as a human effort of recycling organic wastes back into soil to maintain and increase its fertility, for which innumerable methods are being practised. The choice of the method depends upon the availability and the nature of organic wastes and the conditions under which composting takes place. Any process of composting must be flexible and adaptable to every possible set of conditions and must always remain open to new knowledge and actual experience of the practitioners.

WHAT IS VERMICOMPOST

Vermicompost is the excreta (worm castings) of earthworm, which is a natural rich organic soil amendment formed by the decomposition of organic material by the earthworm. It is an odourless, clean, organic material containing adequate quantities of N, P, K and several micronutrients essential for plant growth. Earthworms, necessary to help produce topsoil, eat cow dung or farmyard manure along with other farm wastes and pass it through their body, and in the process convert it into vermicompost. Thus vermicomposting is the process of production of vermicompost through stabilization of organic wastes by earthworm's activity.

Vermicompost improves the structure, texture, aeration and fertility of the soil as well as increases its water-holding capacity. Plants grow stronger and have deeper root systems for better drought-tolerance and disease-resistance. Worms are necessary to help produce topsoil. Vermicompost replaces valuable nutrients taken out of the soil when fruit and vegetables are harvested. It adds beneficial organisms to the soil. The microorganisms and soil fauna help break down organic materials and convert nutrients into a more available food form for plants.

The municipal wastes, the non-toxic solid and liquid wastes of the industries and household garbage can also be converted into vermicompost. Thus in the process of converting garbage into valuable manure, earthworms also help in keeping the environment healthy. Conversion of garbage by earthworms into compost and the multiplication of earthworms are simple processes and can be easily handled by the farmers. Vermicomposting reduces the need for synthetic fertilizers. It is nature's way of recycling.

The foremost advantage of vermicompost is the occurrence of increased level of microbial population, microbial activity, microbial respiration, nitrification, denitrification, enzyme activity and VAM spores. Mucopolysaccharides coated with vermicast act as an important substrate for aerobic, free-living beneficial microbes. So, cellulolytic, lignolytic, nitrifying and nitrogen-fixing microorganisms are found to establish on worm casts (Satchell, 1983). Most of the work reported by Tinuov *et al.* (2001) is related to quantitative estimation of microbes in the vermicast.

The soil mixing process facilitates the movement of ions up and down the soil profile. It should be distinguished clearly from other processes that render ions and other nutrients more readily available to plant rooting system. These processes involve chemical changes that release potential nutrients from bound and unavailable states into forms that can be assimilated by plants. Earthworms may participate in these processes directly and indirectly. An example of a direct effect of earthworms is the mineralization of nitrogen. This element is commonly bound in organic complexes and is not readily available to plants. Passage through the earthworm gut apparently converts this bound nitrogen into more readily available forms such as ammonia, nitrites and nitrates. Considerable quantities of nitrogen locked up in earthworm tissues when these animals are alive, will also be released after their death and decomposition and made available to plants. The relationship between earthworms and microorganisms and the increase in the concentration of Vitamin B12 in soil in the presence of earthworms are examples of the way in which earthworms can participate indirectly in soil enrichment.

Vermicompost is an organic manure (biofertilizer) produced as vermicast by earthworm feeding on biological waste material and plant residues. Vermicompost is a preferred nutrient source for organic farming. It is eco-friendly, non-toxic, consumes low energy input for composting and is a recycled biological product. The process allows for the safe conversion of waste into a valuable nutrient-rich humus fertilizer (vermicompost). Earthworms can be artificially cultured in a brick tank or near the stem/trunk of trees (especially horticultural trees). By feeding these earthworms with biomass, the required quantities of vermicompost can be produced.

SELECTION OF SPECIES FOR VERMICOMPOSTING

Compost worms are big eaters. Under ideal conditions, they are able to consume in excess of their body weight each day, although the general rule-of-thumb is ½ of their body weight per day. They will eat almost anything organic (i.e., of plant or animal origin), but they definitely prefer some foods to others. Manures are the most commonly used worm feedstock, with dairy and beef manures generally considered the best natural food for *Eisenia,* with the possible exception of rabbit manure (Gaddie and Douglas, 1975). The former, being more often available in large quantities, is the feed most often used. The epigeic species have been found to be useful for compost production and the most commonly used species are *Eisenia fetida, Perionyx excavatus* and *Eudrilus eugenia.* These species are fast breeders and feed actively on organic matter high in nitrogen.

CHARACTERISTICS OF MATERIALS FOR VERMICOMPOSTING

Any material that provides a relatively stable habitat for the worms is a good material for vermicomposting. This habitat must have the following characteristics:

1. *Absorbency* Worms breathe through their skins and therefore must have a moist environment in which to live. If a worm's skin dries out, it dies. The bedding must be able to absorb and retain water fairly well if the worms are to thrive.

2. *Bulking potential*   If the material is too dense to begin with, or packs too tightly, then the flow of air is reduced or eliminated. Worms require oxygen to live. Different materials affect the overall porosity of the bedding through a variety of factors, including the range of particle size and shape, the texture, and the strength and rigidity of its structure. The overall effect is referred to as the material's bulking potential.

3. *Content of protein and nitrogen*   Although the worms do consume their bedding as it breaks down, it is very important that this be a slow process. High protein/nitrogen levels can result in rapid degradation and its associated heating, creating inhospitable, often fatal, conditions. Heating can occur safely in the food layers of the vermiculture or vermicomposting system, but not in the bedding.

SELECTION OF MATERIALS FOR VERMICOMPOSTING

Crop residues, tree leaves and animal dung are the basic materials for vermicompost. Agricultural wastes like sugar cane trash, weeds, hedge cuttings, sawdust, coir waste, paddy husk, cattle dung, effluent slurry from biogas plant, excreta of sheep, horse, pig, poultry droppings (in small quantity) and vegetable wastes are ideal food for earthworms. City garbage or even biodegradable organic sludge, a waste product from the effluent treatment plant of any industry, can also be used for feeding worms.

- Some materials make good beddings all by themselves, while others lack one or more of the above characteristics and need to be used in various combinations.
- Shredded paper or cardboard makes an excellent bedding particularly when combined with typical on-farm organic resources such as straw and hay.

Organic producers, however, must be careful to ensure that such materials are not restricted under their organic certification standards.

* Paper or cardboard fibre collected in municipal waste programmes cannot be approved for certification purposes. There may be cases, however, where fibre resources from specific generators could be sourced and approved. This must be considered on a case-by-case basis. Another material in this category is paper-mill sludge (Elvira *et al.*, 1997), which has the high absorbency and small particle size that so well complements the high C : N ratios and good bulking properties of straw, bark, shipped brush or wood shavings. Again, the sludge must be approved if the user has organic certification.

PARAMETERS OF VERMICOMPOSTING

There are a number of other parameters of importance to vermicomposting and vermiculture:

1. *pH*   Worms can survive in a pH range of 5 to 9 (Edwards, 1998). Most experts feel that the worms prefer a pH of 7 or slightly higher. Nova Scotia researchers have found that the range of 7.5 to 8.0 is optimum (George, 2004). In general, the pH of worm beds tends to drop over time. If the food sources are alkaline, the effect is a moderating one, tending to neutral or slightly alkaline pH. If the food source or bedding is acidic (coffee grounds, peat moss) then the pH of the beds can drop well below 7. This can be a problem in terms of the development of pests such as mites. The pH can be adjusted to 7.

2. *Salt level*   Worms are very sensitive to salts, preferring salt contents less than 0.5% (Gunadi *et al.*, 2002). If saltwater seaweed is used as a feed (and worms do like all forms of

seaweed), then it should be rinsed first to wash off the salt left on the surface. Similarly, many types of manure have high soluble salt content (up to 8%). This is not usually a problem when the manure is used as a feed, because the material is usually applied on top, where the worms can avoid it until the salts are leached out over time by watering or precipitation. If manures are to be used as bedding, they can be leached first to reduce the salt content. This is done by simply running water through the material for a period of time (Gaddie and Douglas, 1975). If the manures are pre-composted outdoors, salts will not be a problem.

3. *Urine content* According to Gaddie and Douglas (1975), if the manure is from animals raised or fed off in concrete lots, it will contain excessive urine because the urine cannot drain off into the ground. This manure should be leached before use to remove the urine. Excessive urine will build up dangerous gases in the bedding.

4. *Toxic components* Different feeds can contain a wide variety of potentially toxic components. Some of the more notable are:

- Deworming medicine in manures, particularly horse manure. Most modern deworming medicines break down fairly quickly and are not a problem for worm growers. Nevertheless, if using manure from another farm one's your own, it would be wise to consult the source with regard to the timing of deworming activities, just to be sure. Application of fresh manure from recently dewormed animals could prove costly.

- Detergent cleansers, industrial chemicals and pesticides can often be found in feeds such as sewage or septic sludge, paper-mill sludge, or some food processing wastes.

* Some trees, such as cedar and fir, have high levels of tannins which are naturally occurring substances. They can harm worms and even drive them from the beds. Gunadi *et al.* (2002) point out that pre-composting of wastes can reduce or even eliminate most of these threats. However, pre-composting also reduces the nutrient value of the feed.

VERMICOMPOSTING TECHNOLOGY

In general, there are eight methods of vermicomposting practices that are available. They are,

* Vermicomposting of wastes in field pits
* Vermicomposting of wastes on ground heaps
* Vermicomposting of wastes through tank method
* Vermicomposting on a large/community scale
* Static pile windrows (batch) method
* Top-fed windrows (continuous flow) method
* Wedges (continuous flow) method
* Bin method

VERMICOMPOSTING OF WASTES IN FIELD PITS

Open permanent pits 10 feet long, 3 feet wide and 2 feet deep are constructed under the tree shade, which is about 2 feet above ground to avoid entry of rainwater into the pits (Figure 3.1). Brick walls are constructed above the pit floor and perforated by making 5–6 holes about 10 cm in diameter in the pit wall for aeration. The top of the tank is covered with nylon screen (100 mesh) so that earthworms may not escape from the pits. Partially decomposed dung (dung about 2 months old) is spread on the bottom of the pits to a thickness of about

3–4 cm. This is followed by addition of a layer of litter/residue and dung in the ratio of 1:1 (w/w). A second layer of dung is then applied followed by another layer of litter/crop residue in the same ratio up to a height of 2 feet. Two species of epigeic earthworms, viz., *Eisenia fetida* and *Perionyx excavatus* are inoculated in the pit. The moisture content is maintained at 60–70% throughout the decomposition period. Jute bags are spread uniformly on the surface of the materials to facilitate maintenance of a suitable moisture regime and temperature conditions. It is watered often by sprinkler. The materials are allowed to decompose for 15–20 days to stabilize the temperature, because to reach the mesophilic stage, the process has to pass the thermophilic stage, which comes in about 3 weeks. Earthworms are inoculated in the pit or heap with 10 adult earthworms (1.160.3 g each) per kg of waste material, and a total of 500 worms are added to each pit or heap. The materials are allowed to decompose for 110 days. The forest litter is decomposed much earlier (75–85 days) than farm residue (110–115 days).

Figure 3.1   Production of vermicompost by field pit method

VERMICOMPOSTING OF WASTES ON GROUND HEAPS

Instead of open pits, vermicomposting can be taken up in ground heaps. This method utilizes dome shaped beds (with organic wastes) of dimensions 10 feet length, 3 feet width and 2 feet height (Figure 3.2).

Figure 3.2   Production of vermicompost by heap method

Materials Required

* Farm wastes (straw from wheat, soybean, chickpea, mustard, etc.)
* Fresh dung
* Rock phosphate (Jhabua RP 30–32% P_2O_5) (if P-enriched vermicompost is to be prepared.)
* Wastes : dung ratio (1:1 on dry weight basis)
* Earthworm—1000–1200 adult worms (about 1 kg per quintal of waste material).
* Water—3–5 litres in every week per heap or pit.

Procedure of Vermicomposting

A step-by-step method of preparation of vermiculture bed has to be followed for good results.

1. Select a pit of appropriate dimensions wherever compost is to be prepared.

2. Make a bed of 10 cm height using any of the base materials (coir waste, paddy husk, sugar cane trash, etc.) collected. Add a layer of soil on it. Sprinkle water on it to get a moisture level of 40–45%. The bed should appear wet.

3. Mix the organic waste with cattle dung in equal quantity and pour appropriate quantity of water over it so as to make a homogeneous mixture. Effluent slurry from biogas plant is best used for this. Keep this mixture for two weeks. During this period heating of substrate will take place. Give turning to the material 2–3 times at 4–5 days interval and transfer it to the layer of bedding prepared earlier.

4. Add *Eisenia fetida* species of vermiculture to the bed prepared.

5. Introduce cocoons or worms (if culturing is done for the first time, it is advisable to introduce worms) in the bed at the rate of 2000 worms for 400 kg of feed mix as prepared in the third step. Then, the feed mix is to be spread uniformly on the culture bed. Add 5–10% neem cake in the feed mix. Neem cake in small quantities has beneficial effect on the growth of worms.

6. Cover the bed with gunny cloth. Sprinkle water over the cloth periodically to keep the gunny cloth wet. The worms feed actively on organic matter assimilating only 5–10% of the feed, and the rest is excreted as loose granular mounds of vermicastings on the surface

away from the feed source. Thus the worms will convert the feed mix into vermicasting in 60 days. The vermicompost once formed completely will give the smell of moist soil.

7. Take out the vermicompost and make a heap in sunlight on a plastic sheet. Keep for 1–2 hours. The worms will gather at the bottom of the heap. Remove the vermicompost on the top and the worms settled down at the bottom can be carefully collected for use in the next batch of vermicomposting.

8. Sieve the vermicast and (fine granular materials) vermiwash from the compost heap.

9. The final product (vermicompost) and vermiwash are collected and stored in containers of various sizes.

VERMICOMPOSTING OF WASTES BY TANK METHOD

Vermicompost is prepared in concrete tanks in this method. The size of the tank should be 10 feet length or more depending upon the availability of land and raw materials, 3–5 feet breadth and 3 feet height (Figures 3.3 and 3.4). Suitable plastic tube/ basin structure may also be needed. The floor of the tank should be connected with stones and pieces of bricks.

In this method any type of biodegradable waste can be used. They are,

* Crop residues
* Weed biomass
* Waste from agro-industries
* Vegetable waste
* Leaf litter
* Hotel refuse
* Biodegradable portion of urban and rural wastes.

Figure 3.3   Production of vermicompost by two-tank method

Figure 3.4   Production of vermicompost by four-tank method

Procedure

* Collect the available biowastes and heap under the sun for about 7–10 days and chop if necessary.
* Sprinkle cow dung slurry to the heap.
* Place a thin layer of half-decomposed cow dung (1–2 inches) at the bottom.
* Place the chopped weed biomass and partially decomposed cow dung, layerwise (10–20 cm) in the tank/pot up to a depth of 2½ feet. The biowaste to cow dung ratio should be 60 : 40 on dry weight basis.

* Release about 2–3 kg earthworms per ton of biomass or 100 earthworms per one square feet area. The most suitable species are *Eisenia foetida, Amyanthes diffrigens* and *Eudrilus eugenia.*

* Place a wire net/bamboo net over the tank to protect the earthworms from birds.

* Sprinkle water to maintain 70–80% moisture content.

* Provide a shed over the compost to prevent entry of rainwater and direct sunshine.

* Stop sprinkling water when 90% wastes are decomposed.

* Maturity could be judged visually by observing the formation of the granular compost at the surface of the tank.

* Harvest the vermicompost by scrapping layerwise from the top of the tank and heap under the shed. This will help in separation of earthworms from the compost.

* Sieving may also be done to separate the earthworms and cocoons.

VERMICOMPOSTING ON A LARGE OR COMMUNITY SCALE BY ROOF SHED METHOD

A thatched roof shed preferably open from all sides with unpaved (katcha) floor is erected in east-west direction lengthwise to protect the site from direct sunlight (Figure 3.5). A shed area of 12"× 12" is sufficient to accommodate three vermibeds of 10"× 3 " each having 1"space in between for treatment of 9–12 quintals of waste in a cycle of 40–45 days. The length of the shed can be increased/decreased depending upon the quantity of waste to be treated and availability of space. The height of the thatched roof is kept at least 8 feet from the centre and 6 feet from the sides. The base of the site is raised at least 6 inches above the ground to protect it from flooding during the rains. The vermibeds are laid over the raised ground as per the procedure given below.

Figure 3.5   Production of vermicompost by roof shed method

The site marked for vermibeds on the raised ground is watered and a 4″–6″ layer of any slowly biodegradable agricultural residue such as dried leaves/straw/sugar cane trash etc., is laid over it after soaking with water. This is followed by 1″ layer of vermicompost or farmyard manure.

Earthworms are released on each vermibed at the following rates:

For treatment of cow dung/agriwaste : 1.0 kg per 20 adult worms

For treatment of household garbage : 1.5 kg per 20 adult worms

The frequency and limits of loading the waste can vary as below depending upon the convenience of the user. The appearance of black granular crumbly powder on top of vermibeds indicates harvest stage of the compost. Watering is stopped for at least 5 days at this stage. The earthworms go down and the compost is collected from the top without disturbing the lower layers (vermibed). The first lot of vermicompost is ready for harvesting after 2–2½ months and the subsequent lots can be harvested after every 6 weeks of loading. The vermibed is loaded for the next treatment cycle.

Precautions

* The vermicompost pit should be protected from direct sunlight.
* To maintain moisture level, spray water on the pit as and when required.
* Protect the worms from ants, rats and birds.

STATIC PILE WINDROWS (BATCH)

Static pile windrows (Figure 3.6) are simply piles of mixed bedding and feed (or bedding with feed layered on top) that are inoculated with worms and allowed to stand until the processing is complete. These piles are usually elongated in a windrow style but can also be squares, rectangles, or any other shape that makes sense for the person building them. They should not exceed one metre in height (before settling). Care must be taken to provide a good environment for the worms, so the selection of bedding type and amount is important. After the bedding is supplemented with large quantities of hay and silage and increasing the porosity of the windrows, worm reproduction takes off.

Figure 3.6 Production of vermicompost by static pile windrows (batch)

After two months, the entire bedding material with worms is collected and the vermicompost and worms are separated. The vermicompost may be stored whereas the worms may be inoculated in freshly prepared media for the next batch windrows.

TOP-FED WINDROWS (CONTINUOUS FLOW)

Top-fed windrows (Figure 3.7) are similar to the windrows described above, except that they are not mixed and placed as a batch, but are set up as a continuous-flow operation. This means that the bedding is placed first, then inoculated with worms, and then covered repeatedly with thin (less than 10 cm) layers of food. The worms tend to consume the food/bedding interface and then drop their castings near the bottom of the windrow. A layered windrow is created over time, with the finished product at the bottom, partially consumed bedding in the middle, and the fresher food on top. Layers of new bedding should be added periodically to replace the bedding material gradually consumed by the worms.

Figure 3.7　Production of vermicompost by top-fed windrows (Continuous flow)

The major disadvantages of this system are related to the winter conditions. These windrows require continuous feeding and are difficult, if not impossible, to operate in the winter. In addition, if windrow covers are used, they must be removed and replaced every time the worms are fed, creating extra work for the operator. The advantages of top-feeding have mainly to do with the greater control the operator has over the worms' environment; since the food is added on a regular basis, the operator can easily assess conditions at the same time and modify such things as feeding rate, pH, moisture content, etc., as required. This tends to result in a system with higher efficiency, with greater worm production and reproduction. Harvesting is usually accomplished by removing the top 10–20 cm first, usually with a front-end loader or tractor outfitted with a bucket (Bogdanov, 1996). This material will contain most of the worms and it can be used to seed the next windrow. The remaining material will be mostly vermicompost, with some unprocessed bedding. This can be used as is or screened, with unfinished material put back into the process.

WEDGES (CONTINUOUS FLOW)

The vermicomposting wedge is an interesting variation of the top-fed windrow. An initial stock of worms in bedding is placed inside a corral-type structure (3-sided) more than three feet or one metre in height. The sides of the corral can be concrete, wood, or even bales of hay or straw. Fresh material is added on a regular feeding schedule through the open side, usually by a bucket loader. The worms follow the fresh food over time, leaving the processed material behind. When the material has reached the open end of the corral, the finished material is harvested by removing the back of the corral and scooping the material out with a loader. A 4th side is then put in place and the direction is reversed.

Using this system, the worms do not need to be separated from the vermicompost and the process can be continued indefinitely. During the coldest months, a layer of insulating hay or straw can be placed over the active part of the wedges. The corrals can be any width at all, the only constraint being access to the interior of the piles for monitoring and corrective actions, such as adjustment of moisture content or pH level. A corral width of about 6 feet, with space between and adequate for foot travel, would be ideal. The ideal length will depend on the material being processed, the size of the worm population, and other factors affecting processing times. The sides of the corrals can be made of any material, although insulating value is a consideration. Hay or straw bales will gradually break down over time and be consumed by the worms; as a bale loses its structural integrity, however, it can be added to the contents of the wedge and replaced with a fresh one. The wedge need not have sides at all; in which case it is simply a windrow system where the operator adds feed to one horizontal face, as applied to the top. However, enclosing the sides of the wedge provides a number of benefits, including winter insulation and retention of moisture.

BIN METHOD

Earthworm culturing should be done under shelter to avoid direct sunlight and heavy downpour. Either brick-lined pits, plastic tubs, wooden boxes and earthen pots or even small-sized containers are used for the bin method.

The five basic ingredients needed to start vermicomposting:

1. a container
2. bedding
3. water
4. worms
5. nonfatty kitchen scraps

Container

Collect household food waste for one week (in pounds), and then provide one square foot of surface area per pound. The container depth should be between eight and twelve inches. Bins need to be shallow because the worms feed in the top layers of the bedding. A bin that is too deep is not as efficient and could potentially become an odour problem.

Worm boxes can be purchased or made. Plastic storage containers are convenient and come in a variety of sizes (Figure 3.8). These containers are easily transported and are a nice alternative to heavier wood bins. Many people choose to have several small bins as opposed to one heavier, large wood bin. Small bins work best in homes, apartments and school classrooms. They are easy to tuck under desks, place below kitchen sinks and keep out of the way in laundry rooms.

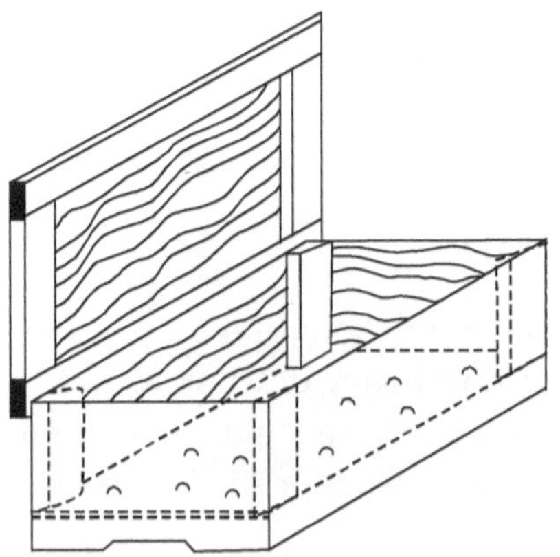

Figure 3.8   Portable bin

Depending on the size of the container, drill 8–12 holes in the bottom for aeration and drainage. A plastic bin may need more drainage. If contents get too wet, drill more holes. Place a tray under the bin to capture excess liquid which can be used as liquid plant fertilizer. The bin needs a cover to conserve moisture and provide darkness for the worms. If the bin is indoors, a sheet of dark plastic or burlap sacking placed loosely on top of the bedding is sufficient as a cover. For outdoor bins, a solid lid is preferable, to keep out unwanted scavengers and rain. Worms need air to live, so the bin should be sufficiently ventilated.

Bedding

The bedding for vermicomposting systems must be able to retain both moisture and air while providing a place for the worms to live. Bedding does not have to be purchased and most of us have plenty of bedding resources in our home, office or school. Here are some suitable sources of bedding. Shredded corrugated cardboard is an excellent bedding. Shredded paper like newspaper and computer paper is easy to find, but may dry out quicker than corrugated cardboard. There is not a problem with the ink from the paper. Commercial worm bedding is available in sporting goods stores, but it is also more expensive.

A 2 × 2 foot box will need between 4 and 6 pounds of dry bedding, a 2 × 3 foot box will take 9 to 14 pounds. No matter what the size, the bin should be $\frac{2}{3}$ filled with prepared bedding. Excess bedding can be dried, stored and used for another time.

Water

Water is needed to moisten the bedding. Place the dry, shredded bedding in a large container and add water until it covers the bedding. Allow the bedding to absorb as much water as possible before putting it in the worm bin. This could take from 2 to 24 hours, depending on the bedding used.

Before putting the bedding in your bin, squeeze the water out from the bedding as much as possible. The bedding should feel like a washcloth. Place the bedding in the bin and fluff. Bedding needs to remain moist. If it is drying out, moisten the paper with water from a spray bottle and dampen the bedding again.

Earthworms

The worms used in vermicomposting are called redworms (*Eisenia fetida*), also known as red wigglers, manure worms, red hybrid or tiger worms. Do not try to use other worms to stock your worm bin because these worms depend on cooler temperatures and an extensive tunnelling system to survive. They will die in worm bin. Redworms prefer temperatures between 13° and 24°C and are suited to living in a worm bin. The temperature of the bedding should not be allowed to get below freezing or above 27°C. The amount of worms needed will depend on the amount of kitchen waste generated per day. One pound of redworms will easily take care of each half-pound of garbage. To add worms to the bin, simply scatter them over the top. The skin on the worm reacts to light and they will immediately work their way down into the bedding to get away from the light.

Kitchen Waste

The kitchen waste fed to worms can come from a variety of sources, including all vegetable and fruit waste (don't be surprised that some seeds may germinate and potato peels with eyes sprout), pasta leftovers, coffee grounds (with filter) and tea bags. Worms may have a problem with garlic and onion skins. Worms have a gizzard like chickens, so fine grit should be added to help the worms digest food. This gritty material includes cornmeal, coffee grounds and/or finely crushed egg shells (dry the shells and then crush). Avoid large amounts of fat, meat scraps or bone. Some sources feel that a small amount of meat and eggs will provide protein to the worms.

It will take time for bacteria to form and your bin can quickly become very smelly if you add too much food, too fast. In the beginning, add a very small amount of gritty material and a small amount of vegetable matter. Gradually increase the amount of food as the bin becomes established. The easiest method is to spread the scraps in a thin layer on top of the bedding. If the bin is kept in a dark place or covered, the worms will come to the surface to eat. You can also pull back a small amount of bedding in the bin and dump in the scraps. Cover the scraps with an inch of bedding. Start at one corner of the bin and bury garbage in a pattern to fill in all the spaces. By the time you get back to the first burying spot, the worms will have composted most of the waste. If you notice odours, cut back on the amount of food or try chopping the food up into smaller pieces. (Note: Citrus does have a strong odour and the peelings seem to last a long time in the bin. Bins seem to be more manageable when there is less fruit and citrus and more of the leafy vegetables).

WORMERY

A wormery is a plastic or wooden container used for vermicomposting common in western countries. The rearing container should be kept in a cool protected place. The wormery should have more surface area than depth because the former ensures more air, quicker composting and thereby quicker feeding and breeding of the worms. Earthworms could be obtained from the garden from moist soil, or from manure pits. There are worm dealers who could also provide worms.

If wooden containers are used, the container should be coated inside with hot paraffin wax. Organic wastes including domestic waste and soil, three times greater in volume that the earthworms are added to the wormery. Use of sandy or clayey soil is to be avoided. 70–80% moisture is maintained in the feed material but it should not be soaked. About 100 worms are added to the containers and the earthworms mate and produce cocoons which produce hatchlings after 3–4 months in the container. The cocoons can be collected from the top layer. Earthworms can live up to 10–15 years but must be fed periodically. So fresh feed (25 kg) should be added every month to the wormery and moisture should be maintained.

The wormery must be kept in a cool place protected from light and at an optimum temperature of 25°C.

Kitchen wastes may include vegetables and fruit waste, used coffee grounds and tea bags. Large amounts of fat, bone and meat should be avoided.

STRUCTURE OF VERMICAST

Not all earthworms produce casts and those that do, cast discontinuously throughout the year. In Britain, *Allolobophora longa* and, to lesser extent, *Lumbricus terrestris* produce definite surface casts mainly during the spring and autumn. In Japan, *Pheretima haipeiensis* casts from April to October. *Allolobophora japonica* does not produce casts at all. The Nigerian *Eudrilus eugenia* produce surface casts but only during the wet season. Earthworm casts vary in size and shape and are often typical of the species producing them. In Britain, *Allolobophora longa* produces small mound-shaped casts, but tropical species may deposit much more conspicuous casts. *Microna chaetids* in South Africa deposit their tower like casts on the side of round cups called Kommetyies, that may be up to 1 m in diameter and 30–100 cm deep.

HARVESTING THE COMPOST

Given the right environment, the worms will go to work to digest the kitchen scraps and bedding faster than any other compost method. The material will pass through the worms' bodies and become "castings." In about 3–4 months, the worms will have digested nearly all the garbage and bedding and the bin will be filled with a rich, black natural fertilizer and soil amendment. Compared to ordinary soil, the worm castings contain five times more nitrogen, seven times more phosphorus and 11 times more potassium. They are rich in humic acids and improve the structure of the soil.

To keep your bin going, you will need to remove the castings from time to time and there are several ways to go about it. One way to do this is to shine a bright light into the bin. The worms are sensitive to light and will move to the lower layers of the bin. Remove the top layer of casting by using your hands or a sieve. Each time you remove some bedding, the worms will be exposed to the light and they will keep migrating down to the bottom of the bin. Pick out any wigglers or worm eggs (small, opaque cocoons) and return them to the bin. Refill the bin with fresh layers of moist bedding and food. Another method of harvesting composts is to push the black, decomposed material to one side of the bin, and fill the other side with new, moist bedding and kitchen scraps. Then wait for several days. The worms will migrate to the freshly filled side of the bin and you can just scoop out the finished compost. Make sure you pick out any wigglers or worm eggs.

PRECAUTIONS FOR STORAGE OF VERMICOMPOST

* Moisture level in the bed should not exceed 40–50%. Waterlogging in the bed leads to an anaerobic condition

and change in pH of medium. This hampers normal activities of worms leading to weight loss and decline in worm biomass and population.

* Temperature of bed should be within the range of 20–30°C.

* Worms should not be injured during handling.

* Bed should be protected from predators like red ants, white ants, centipedes and others like toads, rats, cats, poultry birds and even dogs.

* Frequent observation of culture bed is essential as accumulation of casts retards growth of worms.

* Space is the criterion for growth and establishment of culture. Minimum required space is 2 square metre per 2000 worms with 30–45 cm thick bed.

* Earthworms find it difficult to adapt themselves in new environments hence addition of inoculum as bait from earlier habitat helps in early adaptation to new site of rearing.

CHARACTERIZATION OF VERMICOMPOST

The chemical and biochemical characteristics of vermicompost are presented in Tables 3.1 and 3.2.

Table 3.1   Chemical composition of vermicompost prepared from soybean straw

Parameters	Vermicompost	Vermicompost produced from soybean straw
Ash (%)	51	52
TOC (%)	27	26
C/N ratio	14	13
N (%)	02	01

(Contd.)

Table 3.1 (Continued)

Parameters	Vermicompost	Vermicompost produced from soybean straw
P_2O_5 (%)	02	04
K_2O (%)	01	01
WSC (%)	01	01
Mn (ppm)	500	540
Zn (ppm)	100	100
Cu (ppm)	44	46

Table 3.2 Biochemical characteristics of vermicompost prepared from soybean straw

Parameters	Vermicompost	Vermicompost produced from soybean straw
Total phenol (mg kg^{-1} compost)	98	100
Dehydrogenase (mg TPF kg^{-1} compost hr^{-1})	40	39
Alkaline phosphatase (mg p-nitrophenol kg^{-1} compost hr^{-1})	490	562
Acid phosphatase (mg p-nitrophenol kg^{-1} compost hr^{-1})	398	421

IMPORTANCE OF VERMICOMPOST

Vermicompost appears to be generally superior to conventionally produced compost in a number of important ways.

* Vermicompost is superior to most composts as an inoculant in the production of compost teas.

* Worms have a number of other possible uses on farms, including value as a high-quality animal feed.

* Vermicomposting and vermiculture offer potential to organic farmers as sources of supplemental income.

But working with the worms is a more complicated process than traditional composting. It can be quicker, but to make it so generally requires more labour. It requires more space because worms are surface feeders and will not operate in material more than a metre in depth. It is more vulnerable to environmental pressures, such as freezing conditions and drought. Perhaps most importantly, it requires more start-up resources, either in cash (to buy the worms) or in time and labour (to grow them).

ADVANTAGES OF VERMICOMPOST

Vermicompost is dark brown in colour and is very similar to farmyard manure in colour, use and appearance. It has a fine smell and should have 15–20% moisture.

* Vermicompost is rich in all essential plant nutrients.

* It has an excellent effect on overall plant growth, encourages the growth of new shoots/leaves and improves the quality and shelf life of the produce.

* Vermicompost is free flowing, easy to apply, handle and store and does not have bad odour.

* It improves soil structure, texture, aeration, and water-holding capacity and helps to prevent soil erosion.

* Vermicompost is rich in beneficial microflora such as N_2-fixers, P-solubilizers, cellulose-decomposers, etc. in addition to improving soil environment.

* Vermicompost contains earthworm cocoons and increases the population and activity of earthworms in the soil.

* It neutralizes soil protection.

* It prevents nutrient losses and increases the efficiency of chemical fertilizers.

* Vermicompost is free from pathogens, toxic elements, weed seeds, etc.

* Vermicompost minimizes the incidence of pests and diseases.

* It enhances the decomposition of organic matter in soil.

* It contains valuable vitamins, enzymes and hormones like auxins, gibberellins, etc.

* It helps in the reduction of the noxious qualities of a wide variety of organic waste.

* It helps in protein production for stock food and human food from organic manure.

* It improves the physical structure of the soil.

* It improves the biological properties of the soil (enrichment of microorganisms, addition of plant hormones such as auxins and gibberellic acid, and addition of enzymes, such as phosphates, cellulase, etc.).

* It attracts deep-burrowing earthworms already present in the soil.

* Vermicompost can be used to make compost tea by mixing some vermicompost in water and steeping for a number of hours or days. The resulting liquid is used as a fertilizer.

* Vermicompost fed to poultry stimulates their immune system.

An example of a direct effect of earthworms is the mineralization of nitrogen. This element is commonly bound in organic complexes and is not readily available to plants.

VERMIWASH

During earthworm culture, water is sprinkled over the feed. The excess water and the oily content of earthworms is slowly drained out. The outgoing liquid contains nutrients and beneficial growth hormones for plants. It is a liquid plant growth regulator which contains high amount of enzymes, vitamins, hormones, etc., along with macro- and micronutrients.

Vermiwash Preparation

* A rectangular plastic container is needed. Make a layer of small pieces of bricks and stones of 15–20 cm. Add another layer with soil to a thickness of 15–20 cm.
* Add initially decomposed cow dung to a thickness of 30–40 cm.
* Again add another layer of soil, 3–5 cm thick.
* Then add 100–150 earthworms in the container.
* Finally add a layer of dried leaves which cover the tank for evaporation of water and maintain the temperature.
* At the bottom of the tank, make a hole and put the small container under the hole.
* After one week, the liquid collected in the small container is nothing but vermiwash which has drained from the vermiculture container.
* Everyday, 2–3 litres of vermiwash can be collected.

Application of Vermiwash

One litre of vermiwash is mixed with nine litres of water and it can be sprayed over the plants during the evenings. Another type of application of the vermiwash involves mixing equal amounts of vermiwash and cow urine and adding seven litres of water. The preparation is left undisturbed overnight and then the solution is sprayed over the plant.

There are three methods of applying the vermiwash to plants.

* Direct application to plants.
* Applying the vermiwash to the root tip of transplanted crop.
* Mixing the vermiwash with the seed and then sowing it.

REVIEW QUESTIONS

1. Explain in detail about vermicompost?
2. What are the parameters that affect vermicompost?
3. Describe the methods in vermicomposting technology?
4. List out the advantages of vermicompost?
5. Write a short note on vermiwash and its applications?

4

ROLE OF VERMICOMPOST IN PLANT GROWTH

INTRODUCTION

Soil is the major reservoir for earthworms, microorganisms and also crop production. Vermicompost increases the soil's nutrient-holding capacity and ability to retain moisture, and improves soil structure. The nutrient level in the vermicompost depends upon the species of earthworms. These nutrients will be available to the crops, the year of application.

ROLE OF EARTHWORMS IN SOIL

The degree of importance of earthworms in maintaining soil and crop productivity will vary depending on circumstances. Earthworms are almost always beneficial, when present in the soil. Some soils can be very productive without the presence of earthworms. The worms have sometimes been shown to improve crop growth and yield directly, but more often their activity affects crop growth indirectly through their effects on soil tilth and drainage. Earthworms can have significant impacts on soil properties and processes through their feeding, casting, and burrowing activity. The worms create tunnels in the soil, which can make water and air flow freely as well as enhance root development. The epigeic or surface-dwelling worms create numerous small holes throughout the topsoil, which increases overall porosity and can help improve water and air relationships. Nightcrawlers create large vertical ways, which can greatly increase water infiltration under very intense rainfall or pond conditions. Nightcrawler channels can also increase root proliferation in the subsoil and higher nutrient availability in the cast material that lines portions of the burrow. Earthworm casts, in general, are higher in available nutrients than the surrounding mineral soil, because the organic materials have been partially decomposed during passage through the earthworm gut and the gut microorganisms convert the organic nutrients into more available forms.

Earthworms improve soil structure and tilth. Their casts are an intimate mixture of organic material, mineral soil, etc. The burrowing action of the worms moves soil particles closer together near burrow walls, and the mucus secreted by the worms as they burrow can also help bind the soil particles together. Increased porosity, achieved by residues and soil, are additional ways that earthworms improve soil structure. The mixing of organic materials and nutrients in the soil by earthworms may be an important benefit of earthworms in reduced tillage systems. The earthworms may, in effect, partially replace the work of tillage implements in mixing materials and making them available for subsequent crops. The nutrients released in the natural ecosystem by forest, organisms, leaf litter, etc.

Agricultural practices affect earthworm populations by affecting food supply (location, quality, quantity), mulch protection (affects soil water and temperature), and chemical environment (fertilizers and pesticides). Fertile soils contain highest quantity of earthworms because large amounts of organic materials are continually added to the soil. Fertile fields are usually those in which tillage and the burrow system are left undisturbed. With the addition of plant residues on the soil surface, food supply is continuously available to the earthworms and they live longer and make the soil healthy and wealthy.

IMPORTANCE OF AERATION IN PLANT ROOTS

The stems and leaves of plants are constantly provided with water and dissolved minerals by the efficiency of the root system. Plant roots require air for the process of respiration. Respiration is an essential metabolic process which is fundamental to living organisms, and involves oxygen from the air reacting with stored foods within the plant cells. This releases energy for plant functions especially the uptake of mineral nutrients.

A major limiting factor in the production of quality crops is inadequate drainage and aeration. Aeration is increased or accelerated by soil porosity. The pore spaces in the soil are categorized into either large pores, which normally are filled with air, and the smaller pores, normally filled with water. Roots without adequate aeration and oxygen will result in plants that are weak, that exhibit slow growth and that are predisposed to adverse environmental stresses, such as winter injury, pests and diseases. Carbon dioxide is formed by the respiration of root cells and microorganisms, as well as from the decomposition of organic matter. The plant roots have an environment that is sufficiently porous to avoid the accumulation of carbon dioxide that would lead to the suffocation and eventual death of plant roots. Lack of aeration is caused by poor drainage and waterlogged condition that is a conducive environment for the development of soil-borne pathogens like *Phytophthora* and *Pythium*, responsible for devastating the roots and damping off (Gessert, 1976).

NUTRIENT AVAILABILITY IN VERMICOMPOST

Mineral elements are considered essential to plant growth and development. These elements are involved in plant metabolic functions and the plant cannot complete its life cycle without the element. Usually the plant exhibits visual symptoms indicating a deficiency in a specific nutrient, which normally can be corrected or prevented by supplying that nutrient. Nutrient deficiency can be caused by many plant stress factors. The following terms are commonly used to describe levels of nutrients in plants.

Deficiency   When the concentration of an essential element is low, the yield is affected severely and deficiency symptoms are visible. Extreme deficiencies can result in plant death. Moderate or slight deficiencies may affect the yield but the symptoms may not be visible.

Critical range   A low yield response in the plant indicates that the concentration of nutrient level is low. The optimum nutrient level may be different in different types of plants. The critical range lies between nutrient deficiency and sufficiency.

Sufficiency   The increased concentration of nutrient added to the plants will not increase yield but can increase nutrient absorption by the plant and this is called nutrient sufficiency.

Excessive or toxic   When the concentration of essential or other elements is so high that it reduces plant growth and yields, the element is said to be present in excessive or toxic amounts.

Excessive nutrient concentration can cause an imbalance in other essential nutrients, which also can reduce yield.

Nutrient sufficiency occurs over a wide concentration range, wherein yield is unaffected. Increase in nutrient concentration above the critical range indicates that the plant is absorbing nutrients above that needed level and may not yield product. Elements absorbed in excessive quantities can reduce plant yield directly through toxicity or indirectly by reducing concentration of other nutrients below their critical ranges.

Carbon (C), hydrogen (H) and oxygen (O) are the most abundant required elements in plants. During photosynthesis, green leaves convert CO_2 and H_2O into simple carbohydrates from which amino acids, sugars, proteins, nucleic acids and other organic compounds are synthesized. C, H, O are not considered mineral nutrients. The supply of CO_2 is relatively constant. The supply of H_2O rarely limits photosynthesis directly but does indirectly through the various effects resulting from moisture stress. The various nutrients that are essential to plants and are available in vermicompost are listed below.

Nitrogen

Being a component of proteins, the importance of nitrogen is very great indeed. Nitrogen stimulates vegetative growth of plants and plants supplied with ample amounts of this element are large and succulent and have dark green leaves. On the other hand plants growing in soils deficient in nitrogen are small and pale green and tend to develop fibres to more than ordinary extent. But excessive nitrogen causes replacement of sclerenchyma by collenchyma in the cereal crops and results in the lessening of disease resistance.

The main source of nitrogen for plants is the nitrate or any of the ammoniacal salts present in the soil. The nitrogen of the atmosphere is employed by only a few categories of plants. Vermicompost contains 50% nitrogen.

Sulphur

Sulphates are the chief elements in soil which are essential to plants. The deficiency of sulphur leads to suppression of chlorophyll and the plants turn pale yellow. Sulphur contains certain compounds including some proteins. The characteristic feature of sulphur compounds is the production of odours, e.g. onion, cabbage, radish, mustard, etc. Vermicompost contains 128 ppm of sulphur.

Calcium

Calcium enters into the framework of plants, being the chief constituent of the middle lamellae of plant cells. It stimulates the formation of root hairs. In its absence, starch tends to accumulate in the leaves and therefore it is thought that it has a role to fulfil the function of translocation of foods. Calcium deficiency can be recognized by the chlorotic leaf apices and rolled leaves. Vermicompost contains 2 g of Ca.

Phosphorus

The phosphorus-deficient plants are slow-growing and are stunted on maturity. The dark green colour of the leaves and enhanced development of anthocyanin pigments also characterize phosphorus-deficient plants. Most of the phosphorus of a plant is found in the seeds and fruit. It is a constituent of lecithin, an important lipoid substance commonly present in cells. The plant absorbs large quantities of phosphorus at the time of maturation. This favours healthy root development. Phosphorus is concerned with carbohydrate transformation and also plays an important role in the reduction of nitrates, which is a process very important in protein synthesis. Vermicompost contains 30 per cent of phosphorus.

Potassium

It is found in the form of inorganic salts though to a very minor extent. Potassium is an inducer for the photosynthetic process. It is present in young and active regions of plants such as buds, young leaves and root tips. It has a catalytic influence on synthesis of simple sugars, translocation of carbohydrates, reduction of nitrates, synthesis of proteins and normal cell division. Potassium-deficient plants are stunted at first and eventually dry up and become brown. Their edges and tips appear scorched. Maturation of seeds is somewhat slowed down and there is poor development of mechanical tissues in potassium-deficient plants. 21 per cent potassium is available in vermicompost.

Magnesium

It is indispensable to green plants. Lack of magnesium leads to chlorosis and the chlorotic symptoms appear in the intervenal areas. It occurs in seeds and is more in oil seeds than in starchy ones. It is regarded as essential for oil formation. Magnesium may be related to the formation of nucleoproteins and the process of respiration. It helps to bind the two components of ribose-dimers. Vermicompost contains 2 g of magnesium.

Iron

The chlorotic symptoms caused by iron deficiency appear in the intervenal areas of young leaves, even the smallest veins remaining green. In its absence the young leaves develop a mottled appearance. It is one of the most essential elements of plant nutrition. Plants grown in darkness and exhibiting consequent chlorotic symptoms turn pale green in colour. The iron of haemoglobin in some manner induces the formation of chlorophyll. Iron serves as oxygen carrier and therefore influences respiration. Vermicompost contains 2 ppm of iron.

Manganese

It is one of the trace elements required in very small doses by plants. Its deficiency leads to chlorosis. It plays an important role in the metabolism of plants. Vermicompost contains 2 ppm of manganese.

Boron

It is required by plants in very small amounts. It seems to influence the mobility of calcium inside the plant. In its absence the stem tips die, flower buds do not develop and roots become stunted and thickened in appearance. Boron deficiency in leguminous plants leads to poor development of root nodules. Many physiologic diseases, such as brown heart of turnips, terminal bud breakdown of tobacco, brown core of cabbage and dry rot of beet root have been traced to boron deficiency. Application of borax, ranging from 5 to 50 lbs per acre, cures boron deficiency. Vermicompost contains 2 ppm of boron.

Copper

Copper salts are applied to plants on a long term since they are constituents of fungicides and since copper salts are employed for removing algal growth from fish ponds and swimming pools. It is required for respiration and chlorophyll synthesis. Copper deficiency leads to marked yellowing of leaves. In citrus and many other fruit trees, copper deficiency leads to die-back and also frenching of leaves. Vermicompost contains 2 ppm of copper.

Zinc

Many fruit trees exhibit different kinds of symptoms such as mottled leaf, little leaf, rosette, etc. due to deficiency of zinc. Small fruit and die-back also characterize zinc deficiency. The exact function of zinc is not known but it is thought to act catalytically in oxidation reactions and chlorophyll formation. Vermicompost contains 2 ppm of zinc.

Molybdenum

It is concerned with reduction of nitrate to ammonia prior to amino acid and protein synthesis. Symbiotic and non-symbiotic nitrogen-fixing bacteria like *Rhizobium* and *Azotobacter* seem to require molybdenum. Vermicompost contains 2 ppm of molybdenum.

ANALYSIS OF PLANT GROWTH PARAMETERS BY APPLICATION OF VERMICOMPOST

All parameters of plant particularly their growth and yield attributes need to be measured to analyse the role of vermicompost in plant growth. These parameters include

* Shoot length
* Root length
* Leaf length and width
* Number of leaves, lateral root and branches
* Modulation count

* Fresh weight of stem and root
* Dry weight of stem and root
* Number of pods
* Number of seeds
* Total weight of grains

A comparison of plants grown using vermicompost and that using chemical fertilizers (Figure 4.1) revealed that vermicompost application gave better results.

Figure 4.1 Comparison of vermicompost applied plants and chemical fertilizer applied plants.

APPLICATION OF VERMICOMPOST

The vermicompost could be simply added around the base of each plant. In the case of potted plants, the vermicompost was mixed with the soil. Bacteria and actinomycetes are mainly responsible for the production of vitamin B12 in the soil. The application of vermicompost increases the number of earthworms in the soil, thereby enhancing the flora activity.

ORGANIC FARMING

Organic farming is a method of agriculture that utilizes organic manure and natural methods of plant protection instead of using synthetic fertilizers and pesticides. It is an ecofriendly method that improves soil health and increases the water-holding capacity of the soil, thus making the soil fertile. Organic farming aims at production of quality and safe agricultural products, which contain no chemical residues and which are produced following ecofriendly production methods that restore and maintain soil fertility. Organic farming is becoming increasingly popular worldwide and the global demand for organic products is growing rapidly. It is not only increasing the income of the farmers but also helping in moving towards sustainable agriculture.

Organic farming involves the following components:

* Crop management
* Waste recycling
* Weed management
* Pest control
* Resistance
* Biological chemicals
* Biological disease control
* Bioherbicides
* Bioinsecticides
* Biofungicides

Advantages of Organic Farming

* Organic farming supplies all the nutrients required by the plants.
* It produces optimal conditions in the soil for high yields and good quality crops.
* The growth and physiological activities of the plant are improved.

* The carbon content of organic matter promotes soil aggregation which improve the water-holding capacity and also supplies nutrients to soil microorganisms.

* Organic matter restores the pH but chemical fertilizers create the acidic nature of soil.

* The application of organic manure gives healthy food for man and animals. Organic matter gives more resistant to diseased plants.

Modern agriculture is threatening the natural ecosystem by contaminating soil and water, global warming and ecological degradation. Organic agriculture is the available alternative to conventional agriculture (Laegreid *et al.*,1999) which provides biodiversity and biological activity. But organic farming requires more labour than conventional farming and organic farmers face huge uncertainties. This is due to lack of institutional supports and land tenure. National level efforts are needed to facilitate the successful adoption of organic agriculture by farmers.

REVIEW QUESTIONS

1. What are the nutrients available in vermicompost?
2. Give an account of organic farming.

5

STANDARDS OF HIGH-QUALITY VERMICULTURE PRODUCTS

INTRODUCTION

Live and dead earthworms are commercially available. The dead earthworms are used as fish bait, animal feed, etc. The live earthworms are used in vermiculture units, agricultural lands, horticultural farms and orchards. In addition, ayurvedic medicine manufacturers use them as medicant. But, the commercial success of these vermiculture products depends mainly on their quality. To ensure commercial success and to obtain maximum use from these vermiculture products, these products should be prepared, maintained and stored under some standard conditions.

IMPORTANT PARAMETERS OF PRODUCT QUALITY

To obtain good quality vermiculture products, it is necessary to record some important parameters while processing them. They are

* source, volume and date of receipt of feedstock

* trace ability of feedstock during processing
* temperature and moisture content of material during processing
* pH and electrical conductivity of dissolved salts in processing beds
* screening to 10 mm, bagging or bagging date code on finished product
* trace ability and delivery of finished product

Additional records may be required for some feedstock where special circumstances may apply. The accuracy of record-keeping depends on whether it is a large-scale or small-scale operation.

IMPORTANT TESTS OF PRODUCT QUALITY

To produce good quality vermiculture products, it is essential that they be tested in the laboratory to know their chemical constituents and the nature of the soil. Some of the tests essential to produce quality vermiculture products are

* pH test
* calcium chloride test
* testing the moisture content
* testing the electrical conductivity
* testing the nitrogen content
* testing phosphorus and potassium content
* vermicompost/vermicast stability as an ignition test
* cation exchange for available levels of calcium (Ca), magnesium (Mg), and sodium (Na)
* faecal coliform tests
* heavy metals test

These tests are done to ensure that the end product is odour-free, does not support the growth of the microorganisms and does not produce adverse effects on the growth of the plants.

STANDARDIZATION OF VERMICOMPOST

The end product should be analysed and tested to meet the categorical limits. The test certificate or certificate of analysis of a good quality vermiculture product should have the minimum standards stated in Table 5.1.

Table 5.1   Parameters in the standardization of vermicompost

Parameter	Amount present
C : N ratio	Maximum 25 : 1
SG or bulk density	1.0
Particle size	8 mm
Moisture content (for fresh material)	70 per cent
LOI (loss of weight on ignition)	50–60 per cent
Heavy metals	Nil
FM	Nil
Animal pathogens	Nil
Plant pathogens	Nil
Plant propagules	1per cent
pH	5.0–8.5
EC	1

1. *C : N*   This is the ratio of carbon : nitrogen and gives an indication of the stability of the product. Above 25 : 1, the product will have a noticeable nitrogen demand and may consequently restrict plant growth. But a 20 : 1 ratio gives

sufficient nitrogen to plants and plant growth is accelerated. C : N ratio above 20 : 1 will increase the activity of the microbial population in the vermicast. This reduces the carbon by converting it into carbon dioxide.

2. *Specific gravity (SG)* This is the weight per volume of the combined factors of moisture, mineral and vegetative matter content per grams/litre. Vermicompost should have high organic matter which includes not only vegetative but also the microbiological life like bacteria, actinomycetes and fungi.

3. *Moisture content* Vermicast production mainly depends upon earthworms and microorganisms. If the moisture content is low, the microorganisms may not survive. So, moisture is an important parameter for the survival of microorganisms. The vermicompost may be packed using polythene covers that contain 50% moisture that is necessary for microbial growth. Vermicompost with the required moisture content is very suitable for soil application.

4. *Loss of weight on ignition (LOI)* It is desirable that worm castings should be able to provide some mineral plant nutrition. This mineral portion will be present in the residue after ignition. Sometimes the LOI is maintained by the addition of sawdust.

5. *Foreign matter content (FM)* The vermiculture products to be used in agricultural and horticultural applications should be free from impurities such as plastics, glass particles, mortar, brick chips, gravel, clay, etc. It should not contain any viable seed or plant pathogens.

6. *Heavy metal content* The presence of heavy metals in the vermicompost will affect the growth of plants and animals. For instance heavy metals like lead, mercury, cadmium, chromium, molybdenum, zinc, etc., that enter into food chain would retard the growth of plants and animals.

7. *Animal and plant pathogens*   The vermicompost should not contain animal or plant pathogens. Animal pathogens like *Salmonella* can cause diseases in humans and animals. Plant pathogens are also transferred from one place to another with the help of manure and cause severe environmental damage.

8. *Plant propagules*   A propagule is nothing but a seed or plant shoot. A good quality vermicast should not have any propagules. If they are present they would affect the environment as well as the growth of the plants.

9. *pH*   The most favourable pH is 7. Worm castings should be as close to this as possible.

10. *EC*   Electrical conductivity is the means of determining the salt content. An EC analysis must be carried out to determine the salts present in the vermicompost and its content, especially the chloride content. Many salts are benevolent to plant in low concentrations, for example ammonium sulphate. If the worm castings show a high EC, then it must be disqualified from being sold.

11. *General appearance*   Compost worms produce commercial quantities of castings, which are produced as aggregates. Aggregates are the building blocks of erosion-resistant soils (Edwards, 1998). Aggregates remain soft and spongy whether damp or dry. These products should be dark grey to black in colour in small aggregate form and it should contain 50% moisture. Compressed in the bag, it should exhibit a clear elastic property.

PROCESSING OF THE VERMICOMPOST

After setting the standard parameters for producing quality vermicompost, it is processed for commercial production. The following steps should be considered while processing the vermicompost.

1. Every ten days remove the vermicompost (earthworm excreta) from the heap.

2. Put all the contents in the floor. After 24 hours, the earthworms settle at the bottom; then the earthworms and vermicompost could be separated easily.

3. The vermicompost is sieved with the help of one-mm sieve and spread in a thin layer for air-drying.

4. The remaining unsieved material is retained as partially decomposed organic matter for breakdown by earthworms.

5. In one month all the partially decomposed organic matter gets converted into vermicompost.

6. After 10 days of drying, vermicompost is collected and put in a heap. This is done to remove the remaining earthworms. For this purpose about half a kilogram of cattle dung is added inside the heap. In about two days all remaining earthworms settle in the cattle dung. This cattle dung is removed and placed on the recharged matter.

The chambers used for processing vermicompost should be recharged with partially decomposed organic matter and the earthworms as early as possible.

Plants fed only vermicompost or vermicompost slurry may suffer from nutrient shortage (particularly N and P) in the short-term (6–18 months). Moreover, a good manure should contain high water-holding capacity. Dried vermicompost that does not readily re-wet will further slow down the mineralization process, extending the time frame of any nutrient shortage. So, a wetter must be added to avoid drying out, and the vermicompost must be diluted with other media if root growth inhibition is to be avoided. The required plant nutrients are supplied by vermicompost.

PACKING AND SALE OF VERMICOMPOST AND EXCESS EARTHWORMS

After the commercial production process, the vermicompost and the earthworms are packed and marketed to the potential buyers. Before packing, the vermicompost is dried in open environment. Then they are packed in 1–5 kilogram packs in air-tight polythene bags. Vermicompost can be sold at the rate of Rs. 3/kg. Excess earthworms along with a little vermicompost can be sold at Rs. 300/1000 earthworms.

NUTRIENT CONTENT OF VERMICOMPOST

A good quality vermicompost should have a well-balanced nutrient content. According to Radha, D. Kale (1998) vermicompost contains many nutrients as listed in Table 5.2.

Table 5.2 Nutrient content of vermicompost

Nutrients	Percentage or ppm
Organic carbon	55 per cent
Total nitrogen	50 per cent
Available phosphorus	30 per cent
Available potassium	21 per cent
Calcium and magnesium	2 gm
Copper	2 ppm
Iron	2 ppm
Zinc	5 ppm
Sulphur	128 ppm

ECONOMIC VIABILITY OF VERMICOMPOST

The fact that 100 kilograms of organic matter is converted into almost 100 kilograms of vermicompost and that excess earthworms are produced reveal that vermicomposting is a profitable agriculture-related business.

a) *Expenditure*   The various costs involved in the production of vermicompost are listed in Table 5.3.

Table 5.3   Expenditure items and cost of vermicompost

Expenditure item	Cost (Rs.)
Organic matter (parthenium weed) 100 kg	40
Cattle dung 16 kg	16
Mixing of cattle dung with water with 100 kg organic matter	20
Filling of partially decomposed organic matter in four 3 feet × 3 feet × 3 feet chambers and putting 4000 earthworms	15
Cost of 4000 earthworms	1,200
Processing cost of 100 kg vermicompost	40
Packing of 100 kg vermicompost	20
Total cost of 100 kg vermicompost	1,351

The total cost involved to produce 100 kg of vermicompost is Rs. 1351.

b) *Earnings*   The income earned from selling 100 kg of vermicompost is Rs. 300.

The income earned from selling the additional earthworms produced (8000 nos.) is Rs. 2400.

Therefore the total income earned is Rs. 2700

Hence, the net profit is 2700 − 1351 = Rs.1349

So, it is apparent that vermicompost agro-industry is quite profitable.

REVIEW QUESTIONS

1. How is vermicompost processed?
2. Explain about the economic viability of vermicompost?
3. How is the end product of vermiculture tested for standardization?

VERMICULTURE FOR WASTE REDUCTION

INTRODUCTION

Nowadays, to meet the increasing demands in the market, there is a growing competition among the agriculturists and the horticulturists for producing the best varieties of vegetables, fruits and ornamental flowers. Chemical fertilizers added to plants to produce vegetables and fruits cause serious health problems. So adopting vermiculture would be a solution to these problems. The advantage of this technology is that the treated waste acts as a high-quality soil conditioner that retains the characteristics of the soil. Research efforts of the scientists on vermiculture technology has helped to produce high quality solid and liquid biofertilizer products that are ecofriendly and cost-effective. It is possible that the chemical fertilizers would be replaced by biofertilizers in the future.

CHEMICAL FERTILIZERS

Chemical fertilizers contain only 5 or 6 elements in high concentration (N, P, K are the prime elements). Application of chemical fertilizer has affected living beings, soil organisms and the environment. The continuous use of large quantities of chemical fertilizers to increase the yield and use of improved crop varieties and pesticides cause several hazards such as heavy loss of macronutrients (Prasad and Singh, 1981); deficiency of micronutrients (Kanwar and Radhawa, 1978); nutrient imbalance (Singh *et al.*, 1989) and reduction in organic matter content (Padmaja *et al.*, 1996). As a result soil gets deteriorated, which in turn affects the plant productivity. The development of organic farming could maintain both soil fertility and plant productivity. In organic farming, organic manures and biofertilizers are mainly used. In the era of sustainable organic farming, vermicomposting for the production of organic manure and the use of vermicompost in agricultural lands is a breakthrough in the field of agriculture. Chemical fertilizers have been responsible for deterioration of soil friability (crumbliness), and destruction of beneficial soil life such as earthworms, bacteria and micro arthropods. Moreover plants grown on artificial fertilizers have lower nutrient value than their organically grown counterparts.

The most harmful impact of chemical fertilizers is that these fertilizers leach into the subsoil and permanently contaminate the soil and the groundwater. Moreover they also enter the food chain and create health hazards.

BIOFERTILIZER

Microorganisms are added to soil to enhance the fertility of the soil. This is due to the ability of many of these microbes to fix atmospheric free nitrogen which is abundantly available. They are useful in supplementing the usual application of chemical nitrogen fertilizers and help enriching the soil. Cultivation of leguminous species as a means of enriching the fertility of soil is long known due to their ability of fixing atmospheric nitrogen with the help of the bacterium *Rhizobium* present in their root nodules. Cyanobacteria are involved in N_2 fixation. Some of these microorganisms are symbiotic and some are free living in the soil. Their effective use help avoiding chemical fertilizers and prevent pollution. More than half a dozen species of *Rhizobium* are associated with the root nodules of several legume plants, while *Azospirillum* is associated with the root hairs of grasses and *Frankia* is associated with a number of other shrubby and woody plant root systems including *Alnus* and *Casuarina.* Other bacterial genera endowed with such a capacity include species of *Klebsiella, Azotobacter, Clostridium, Rhizobium* and *Actinomycetes* which are free-living. Among the algal group, the two well-known blue-green algae are *Anabaena* and *Nostoc,* which are free-living in soil. But the algae, *Azolla* occurs in symbiotic association in the leaf-pockets of the free-floating aquatic fern, *Azolla.* Now *Azolla* is commonly used as an organic input in rice fields (Kannaiyan, 1987). It is a low-cost biofertilizer and can add up to 40–60 kg N/hectare/crop (Cannon, 1978; Brill, 1980).

Besides the use of microbial biofertilizers, there is a possibility of improving the soil texture with the use of organic fertilizers obtained from the huge organic wastes and through bioprocessing of huge biomass of aquatic weeds, which are normally considered as obnoxious (Postgate, 1982). In India, one national centre and several regional centres have been established to develop and improve the technology of biofertilizer production (Hollander, 1977). For biofertilizer production earthworms are the ideal managers, whom man can employ to maximize growth of aerobic bacteria for waste processing.

WASTE REDUCTION

Waste is not just created when consumers throw items away. Manufacturing any product needs raw material. But the raw materials are not fully utilized in any production process and some waste is produced. Also, some amount of raw material is wasted during transportation, handling, storage, etc. At every moment we are releasing unused and unwanted material in the environment. Using vermiculture technology, reuse of waste materials with minimum loss is possible. Ultimately, less material has to be recycled or sent to landfills or waste combustion facilities. Selecting non-hazardous or less hazardous items is another important component of source reduction. Using less hazardous alternatives for certain items (e.g. cleaning products and pesticides), sharing products that contain hazardous chemicals instead of throwing out leftovers, reading label directions carefully, and using the smallest amount necessary are ways to reduce waste toxicity. Minimum waste means increase in saving for communities, businesses, institutions, schools and individual consumers. Using reusable products reduces waste and pollution.

a) Solid Wastes

All over the world, industrialization, urbanization and intensive agricultural practices have led to an increase in the generation of wastes from various sources. The complex structural composition of the waste resists their breakdown. So natural decomposition will become a slow process and result in the accumulation of these wastes. The main reason for environmental pollution is the wastes produced due to human activities. Ever increasing population, intensive agricultural and chemical processes and the modern life style are mainly responsible for the constant increase in the production of solid wastes. The wastes generated from houses, hotels, hospitals, industries, etc. are hazardous to environment when they are left untreated. But, these wastes are rich in nutrients required for the crops. If they are processed in a proper way, they can become a good source of manure. Usually, the solid wastes thrown out in public places or roadsides are collected and used to fill up low-lying areas, which again pollutes the environment. To prevent such hazardous activities and also to derive beneficial products from the wastes, several techniques are available. These techniques are mainly based on the reuse, recycle and recovery of resources from the organic wastes. Conventional agriculture depends largely upon the external application of inorganic fertilizers to soils for maximizing the crop productivity. As a result, there is an increasing gap between depletion and replication of nutrients combined with decreasing biological activities. Municipal solid wastes generated in urban areas cause health hazards in most parts of the world. On an aggregate, it has been found that over 50% of the wastes are made up of biodegradable organic wastes.

b) Biomedical Wastes

Biomedical waste is any solid or liquid waste, which may present a threat of infection to humans. Biomedical waste is generated from health care establishments such as hospitals, blood banks, laboratories and research institutes, veterinary hospitals, agriculture and food-processing industries, etc. Hospital waste includes non-liquid tissue, body parts, blood and blood products, and body fluids from humans and other primates, laboratory and veterinary wastes which contain human disease-causing agents, and discarded sharps. Discarded sharps include needles, syringes, blades, scalpels, slides, broken glass, blood-stained clothes, cotton, etc.

Biomedical waste management is a special case wherein the hazards and risks exist not just for the generators and operators but also for the general community. Generally biomedical waste is classified into infectious waste and non-infectious waste, though the proportion of infectious waste is low. If infectious waste was not disposed off scientifically, it could contaminate non-infectious waste. With the ever-growing population and the increase in number of hospitals and other medical laboratories, the problems of pollution from biomedical waste are also increasing simultaneously.

The treatment of these wastes depends upon the nature of the waste, technology that is technically and economically viable and environmentally safe. Vermiculture technology is an economically viable and environmentally safe technology for the process of treating biomedical wastes.

c) Domestic Wastes

Population growth, rapid urbanization and other development activities during the past few decades have been responsible for environment pollution and resources degradation. Rapid urbanization has seriously created the problem of municipal or domestic waste management. Municipal solid waste consists of kitchen waste, fruit and food waste, glass, paper, plastic, metal, rags, packing material, etc.

A large volume of domestic solid waste is generated in both urban as well as rural areas. It includes organic and inorganic waste. These wastes are dumped in open spaces causing land, air and water pollution. The dumping sites are not properly managed nor have been planted with suitable plant species to help in quick degradation of solid waste by way of creating an environment conducive for the growth of microorganisms besides providing greenery. Appropriate post-dumping practices are also seldom performed causing perpetual problem of air and water pollution.

The physical and chemical composition of the urban waste is given in Tables 6.1 and 6.2.

Table 6.1   Physical composition of domestic waste (Density = 230–550 kg/m³)

Category	Item	Per cent
Recyclable material	Paper, plastic, rags	4–6
	Leather, rubber, synthetic	2–3
	Glass, ceramics	1
	Metal	2
Compostable material	Food articles, fodder, dung, night soil, leaves, organic material	50–70
Inert material	Ash, dust, sand, building material	30–60
Moisture		50–90

Table 6.2   Chemical composition of domestic waste (Calorific value = 800–1010 kcal/kg)

Item	Percentage
Nitrogen	0.56–0.71
Phosphorus	0.52–0.82
Potassium	0.52–0.83
C/N	21–30

In rural areas, the household waste and the cattle dung are generally collected and dumped at the dumping sites located outside the village. These wastes are mostly used as manure for agricultural products. The cattle dung is also used to make cattle-dung cakes that are used as an alternative to wood used for cooking purposes. The characteristic feature of these solid wastes collected from the rural areas is that they are free from glass, metal or other non-biodegradable material. Though these wastes are disposed from the dumping sites in a regular manner, using vermiculture technology for the disposal of these wastes would yield better results.

d) Industrial Wastes

The storage of industrial solid waste is often one of the most neglected areas in the operation of a firm. Very little attention is paid to proper storage, and heaps of mixed waste piled against a wall or dumped in open grounds are a common sight in many factories. Concrete bays or disused drums are also often used for storage, whereas the sludges originating from these industries are let to the drainage without any treatment.

Waste is regarded as an unwanted product by firms and there is no waste-control and waste-maintenance department in many industries. Transportation of industrial waste in metropolitan areas of developing countries is generally not done by purpose-built vehicles such as skip-carrying lorries, but done by open trucks. Moreover, the wastes are not covered during transportation. There are no special arrangements made for transporting hazardous wastes; they are usually collected together with the other wastes. Generally, there is little control over either the types of firms engaged in carrying hazardous waste or the vehicles used.

Industrial waste, whilst presenting the same disposal problems as domestic waste, also contains hazardous waste, thereby exacerbating the difficulties of disposal. In the past there has been little control over the disposal of industrial wastes; indeed, it has only been during the last decade that even developed countries have brought in legislation to curb the uncontrolled and environmentally unaccept practices that were widespread in disposing the wastes. These wastes could be treated with microorganisms in an environmentally safe manner.

Tapioca or cassava (*Manihot esculenta* Crantz) is an important staple food cum industrial cash crop of the tropics. It plays a vital role in the economy of the country and is gaining significant importance in the recent years as one of the major sources of sago, starch, glucose, ethanol, etc. *Thippi* is generated at the rate of 15–20 per cent per ton of the tapioca tubers processed during the manufacture of starch and sago. The organic and volatile matters present in the thippi are nearly more than two-thirds of its weight and may cause serious problems especially due to growth of undesirable microflora in the soil. Generally *thippi* is allowed to stay in wet condition near the rasping section and in due course they start developing foul odour due to microbial action. So, partially decomposed thippi can be added to pressmud and then used for vermiculture. It will give good manure for agricultural use. So, the industrial wastes can be converted into vermicompost that would reduce the wastes in the environment.

WASTE DEGRADATION BY VERMICOMPOSTING

The total annual waste generated in India in the form of municipal solid waste is 25 million tons, agriculture waste residues 320 million tons, cattle manure 210 million tons and poultry manure is 3.3 million tons. Traditionally the solid waste management practices involve collection and transportation to far off low-level dumping sites. This leads to fowl smelling area, spread of diseases and mosquito breeding grounds. The other option is composting which involves the dumping of waste into a pit that is covered with soil on the top. The bioconversion of waste to farmyard manure (FYM) by this method takes about 6 months. Figure 6.1 shows the plant used for the large-scale production of vermicompost from industrial wastes.

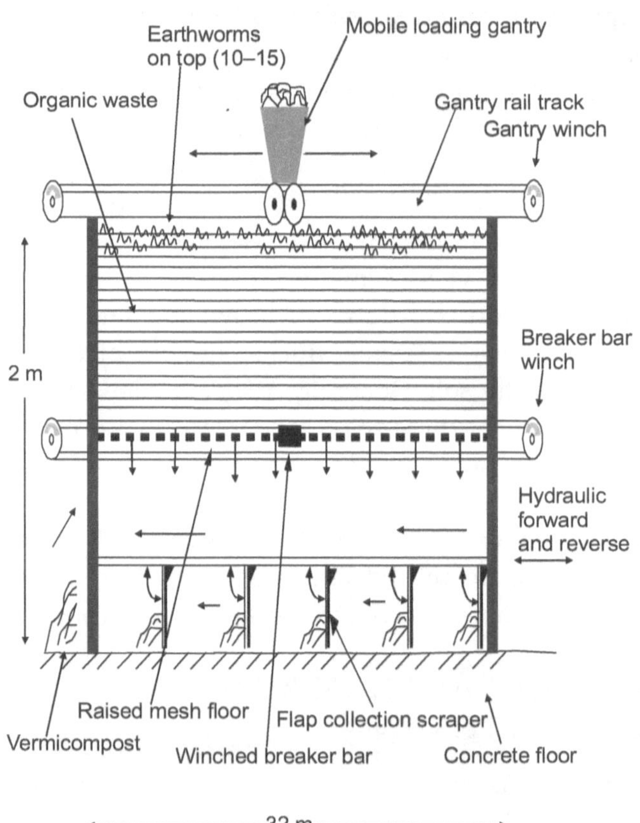

Figure 6.1 Large-scale vermicomposting plant

There is a tremendous scope to convert the biodegradable waste into organic manure through vermiculture biotechnology or vermicomposting. Vermiculture biotechnology denotes the use of earthworms as natural bioreactors for efficient biodegradation of organic solid wastes. The bacterial species such as *Klebsiella pneumoniae, Morganella morgoni, E. coli, Bacillus subtilis* are present in the vermicast. It may be inferred that these beneficial nitrogen-fixing microbes can help the plant to a greater extent and also play a vital role in the process of organic waste degradation. It may have beneficial effect on plant growth as has been stated by Jeffries (1987), and Reddell and Spain (1991).

REVIEW QUESTIONS

1. Write a short note on biofertilizer?
2. Describe the role of vermiculture in waste reduction?

GLOSSARY

Anecic   The type of earthworms that build permanent vertical burrows that extend through the upper soil layer. They make burrows as deep as 4–6 feet.

Biofertilizer   A fertilizer mainly composed of soil microbes which are beneficial for crop production. They are useful in supplementing the usual chemical nitrogen fertilizers and help in enriching the soil.

Chemical fertilizer   A fertilizer mainly prepared from chemicals. The most harmful impact of chemical fertilizer is that these fertilizers leach the subsoil and permanently pollute the soil and the ground water.

Clitellum   A thick belt of smooth skin girdle muscle around the body of an earthworm normally occurring around the 14th and 15th segment. It is an identifying feature for mature earthworms.

Cocoon   The mucus sheath produced by the clitellum of mature earthworms. This sheath containing nutritive fluid slides forwards, collects the eggs and sperms from the girdal opening of the worm. The sheath finally slides off from the head of the worm and its ends are sealed and is now called the cocoon. The worm deposits the cocoon in the topsoil. The cocoons can be identified easily in the morning. They are easy to spot in the worm bed.

Compost   The humuslike substance produced by the biological process in which microorganisms, mainly fungi and bacteria, convert degradable organic wastes into humuslike substance. It helps the soil to retain more of the plant nutrients. It is a slow process when compared to vermicompost process.

Composting worms   The earthworms which are suitable for the production of manure.

Critical range   The critical range of nutrient level that may vary from plant to plant and from species to species, and that lies between nutrient deficiency and nutrient sufficiency.

Earthworm   A free living soil animal which is present in dark places in mud. It is of great economic value to mankind because it improves the soil quality by its burrowing action. It is often called as farmer's friend.

Endogeic   The earthworms that live in a burrow system and that rarely come to the surface. These worm species help to incorporate mineral matter in the topsoil layer and mix the soil through their movement and feeding habits.

Epigeic   The type of earthworms living in the topsoil layer where they feed on decaying organic matter. This type of earthworms are mainly used for vermiculture and vermicomposting practices.

Hatchling   Small earthworms coming out from the cocoons after the period of incubation.

Light-sensitive cells   Cells present in the skin of earthworm and used to detect the light changes in the environment. The cells are very sensitive to touch and chemicals.

Micro-irrigation drip emitter   A small tank which contains a liquid fertilizer that can be pumped to each individual plant uniformly and directly.

Monoculture   The culture of a single species of earthworms.

Nightcrawlers   The earthworms that can be collected at night during spring and rainy season.

Nodules   Structures present in the roots of leguminous plants which aid in fixing the atmospheric nitrogen. The healthy nodule is pink in colour.

Noncomposting worms   The earthworms that are not suitable for the production of manure. It is mainly available in gardens.

Nutrient magnification   Earthworms not only enhance the activities of soil microorganisms but, by grinding the soil they also speed up the mineralization process, by which nutrients are released. In this way, nutrient magnification can take place.

Organic farming   A method of agriculture that utilizes organic manure and natural methods of plant protection instead of using synthetic fertilizers and pesticides.

Polyculture   The culture of many species of earthworms in a single container.

Porphyrin   The colour pigment which is present in the integument of earthworms. It protects the worm from the injurious effects of bright light.

Setae   Locomotory organs in earthworms. Setae are present in all body segments except first and last segment.

Soil porosity   Soil contains large and small pores. The large pores are filled with water. Soil porosity is important for aeration of the soil.

Vermicast   A process in which the digested food material passed out of the earthworm through the anus in a circular contraction of muscles starting from anterior end to rectum. Vermicast is in the form of pellets and is a dark brown/black humuslike coarse material, soft in feel and free from contamination. The mucus coated on the vermicast increases aeration in the soil, provides excellent water retention properties and improves drainage in heavy soils.

Vermiculture   The artificial rearing or cultivation of earthworms for the production of vermicompost which is rich in humus.

Vermiwash   The brown coloured liquid fertilizer collected after passage of water through a column of earthworms in partially decomposed organic matter. It can be used as a foliar spray for all crops.

Wormery   A container (either wood or plastic) consisting of composting worms and used for the production of vermicompost.

References

Arancon, N.Q., Edwards, C.A., Bierman, P., Metzger, J.D., Lee, S. and Welch, C. (2003). "Effects of vermicomposts on growth and marketable fruits of field-grown tomatoes, peppers and strawberries." *Pedobiologia.* 47: 731–735.

Asthana and Chaturvedi. (1999). "A little inputs needed." *The Hindu Survey of Indian Agriculture.* pp. 61–65.

Atlavinyte, O. and Daclulyte, J. (1969). "The effect of earthworms on the accumulation of vitamin B12 in soil." *Pedobiologia.* 9: 165–170.

Bano, K., Kale, R.D. and Gajanan, G.N. (1987). "Culturing of earthworm *Eudrilus eugenia* for cast production and assessment of worm cast as biofertilizer." *J.Soil.Biol.Ecol.* 7(2): 98–104.

Bogdanov, Peter. (1996). *Commercial Vermiculture: How to Build a Thriving Business in Redworms.* Vermico Press, Oregon. pp. 83.

Bouche, M.B. (1977). "Strategies lombriciennes." In: *Soil Organisms as Components of Ecosystems.* Lohm, U. and Persson, T. (eds.). *Ecol. Bull.* (Stockholm). 25: 122–132.

Brill, W.J. (1980). "Biochemical genetics of nitrogen fixation." *Microbiol. Rev.* 44: 444–467.

Cannon, F.C. (1978). "Cloning of *Klebsiella* nitrogen fixation gene (*Nif*)." *Genetics.* Boyer, H.W. and Nicosia, S. (eds.). North Holland Biomed. Press, Amsterdam.

Darwin, C.R. (1881). *The Formation of Vegetable Mould through the Action of Worms with Observations on their Habits.* Murray, London.

Dash, H.K., Beura, B.N. and Dash, M.C. (1986). "Gut load, transit time, gut microflora and turnover of soil, plant and fungal material by some tropical earthworms." *Pedobiologia.* 29: 13–20.

Day, G.M. (1950). "The influence of earthworms on soil microorganisms." *Soil.Sci.* 69: 175–184.

Domínguez, J. (2004). "State of the art and new perspectives on vermicomposting research." *Earthworm Ecology.* Edwards, C.A. (ed.). CRC Press, Boca Raton, FL, USA. pp. 401–424.

Edwards, C.A. (1998). "The use of earthworms in the breakdown and management of organic wastes." In: *Earthworm Ecology.* Edwards, C.A.(ed.). St. Lucie Press, Boca Raton. pp. 327–354.

Edwards, C.A. (1998). "The use of earthworms in the breakdown and management of organic wastes." *Earthworm Ecology.* Edwards, C.A. (ed.). CRC Press, Boca Raton, FL, USA. pp. 21–31.

Edwards, C.A. and Burrows, I. (1988). *The Potential of Earthworm Composts as Plant Growth Media.* SPB Academic Publishing, The Hague, The Netherlands.

Elvira, C., Sampedro, L., Domonguez, J. and Mato, S. (1997). "Vermicomposting of wastewater sludge from paperpulp industry with nitrogen rich materials." In: *Soil Biol. Biochem.* 29 (3/4). pp. 759–762.

Gaddie, R.E. and Donald, E. Douglas. (1975). "Earthworms for ecology and profit." *Scientific Earthworm Farming.* Vol. 1. Bookworm Publishing Company, CA. pp. 180.

Gavrilov, K. (1962). *Role of Earthworms in the Enrichment of Soil by Biologically Active Substances.* Voprosy Ekologi Vysshaya Shkola, Moscow. 7: 34.

Georg. (2004). "Feasibility of developing the organic and transitional farm market for processing municipal and farm organic wastes using large-scale vermicomposting." Good Earth Organic Resources Group, Halifax, Nova Scotia.

Gessert, G. (1976). "Measuring air space and water-holding capacity." *Ornamentals Northwest.* pp. 3: 59–60.

Grundon, N.J. (1980). "Effectiveness of soil dressing and foliar sprays of copper sulphate in correcting copper deficiency of wheat (*Triticum aestivum*) in Queensland." *Australian Journal of Experimental Agriculture and Animal Husbandry.* 20: 717–723.

Gunadi, Bintoro, Charles Blount and Clive, A. Edwards. (2002). "The growth and fecundity of *Eisenia fetida* (Savigny) in cattle solids pre-composted for different periods." *Pedobiologia.* 46: 15–23.

Hervas, L., Mazueles, C., Senesi, N. and Saizjimenez. (1989). "Chemical and physico-chemical and biological characterization of vermicompost and their humic acid fractions." *The Science of the Total Environment.* 81/82. 543–550.

Hollander, A. (ed.). (1977). *Genetic Engineering for Nitrogen Fixation.* Plenum Press, New York.

Jeffries, P. (1987). "Use of mycorrhizae in agriculture." *Crit. Rev. Biotech.* 5: 319–357.

Julka, J.M. (1993). "Earthworm resources of India and their utilization in vermiculture." In: *Earthworm Resources and Vermiculture.* Zoological Survey of India, Calcutta. pp. 51–56.

Kale, K.E. (1998). *Earthworm: Cinderella of Organic Farming.* Prism Books Pvt. Ltd. Bangalore. p. 88.

Kale, R.D. (1994). Consolidated Technical Report of the ADHDC Scheme on Promotion of Vermicomposting for Production of Organic Fertilizer. G.K.V.K, Bangalore, India.

Kannaiyan, S. (1987). "Azolla Biotechnology." Technical Bull, TN Agri. Univ. Coimbatore, India.

Kanwar, J.S. and Radhawa, N.S. (1978). "Micronutrient research in soil and plants in India (A Review)." Tech. Bull. 50. ICAR, New Delhi.

Laegreid, M., Beckman, O. C. and Kaarstad, O. (1999). *Agriculture, Fertilizers and the Environment.* CABI Pub. and Norsk Hydro. ASA, Oxon, UK.

Mba, C. (1983). "Utilization of *Eudrilus eugenia* for disposal of cassava peel. *Earthworm Ecology: From Darwin to Vermiculture.* Satchell, J.E. (ed.). Chapman and Hall, London. pp. 315–321.

Myers, Ruth. (1969). *The ABCs of the Earthworm Business.* Shields Publications, Eagle River, Wisconsin, USA. pp. 64.

Owen, W.L. (1954). "Dewaxed, retreated, refining filter pressmud as a plant food and plant growth stimulant." *Sugar.* 49 (3): 40–42.

Padmaja, P., Prabha Kumari, P., Jiji, T. and Ushakumari, K. (1996). "Vermitechnology for composting biowastes." Nat. Sem.Org. Fmg and Sust. Agric. Bangalore. Veeresh, G.K and Shivashankar, K. (eds.). p. 36.

Postgate, J.B. (1982). *Biological Nitrogen Fixation Fundamentals.* Phil. Trans. Royal Soc. (B), London. 296: 375–385.

Prasad, B and Singh, R.P. (1981). "Accumulation and decline of available nutrients with long term use of fertilizers manure and lime on multiple cropped lands." *Indian J. Agric. Sci.* 51: 108–111.

Prasad, R. and Power, J.F. (1997). *Soil Fertility Management for Sustainable Agriculture.* CRC-Lewis, Boca Raton, USA.

Reddell, P. and Spain, A.V. (1991). "Earthworms as vectors of viable propagules of mycorrhizae fungi." *Soil.Biol.Biochem.* 23 (8): 767–774.

Satchell, J.E. (1983). "Earthworm microbiology." In: *Earthworm Ecology From Darwin to Vermiculture.* Satchell J.E. (ed.). Chapman and Hall, Cambridge. pp. 351–364.

Senapati, B.K. and Dash, M.C. (1984). "Functional role of earthworms in the decomposer subsystem." *Tropical Ecology.* 25 (1): 52–72.

Singh, A.P., Sakal, R. and Sinha, R.B. (1989). "Effect of changing cropping pattern and fertility levels on crop yields and micronutrients status of soil after five cycles of crop rotation." *Ann. Agric. Res.* 10(4): 361–367.

Solayappan, A., Muthukumaraswamy, R., Revathi, G., Mala, S.R. and Vadivelu, M. (1997). "Recycling of sugar cane wastes: A biotechnological approach." Proc.Sem.on recycling of Biomass and Industrial by-products of Sugar cane. Cuddalore. pp.21–28.

Springett, J.A. and Syers, J.K. (1979). "The effect of earthworm casts on rye grass seedlings." In: Proc.2nd Australian Conf. Grassl. Invert. Ecol. Crosby, T.K. and Pottinger, R.P. (eds.). Government Printer, Wellington pp. 44–47.

Timothy, B. Parkin. and Edwin, C. Berry. (1999). "Microbial nitrogen transformations in earthworm burrows." *Soil Biol. and Bio. Chem.* 31: 1765–1771.

Tiunov, A.V., Dobrovolskaya, T.G. and Polyanskaya, L.M. (2001). "Microbial complex associated with inhabited and abandoned burrows of *Lumbricus terrestris* earthworm in soddy–Podzolic soil." *Eurasian Soil Sc.* 34 (5): 525–529.

Related Websites

en.wikipedia.org/wiki/

www.erfindia.org/

www.imdb.com/title/tt0111948/

www.sarep.ucdavis.edu/worms/

www.backyardnature.net/earthworm.html/

www.bae.ncsu.edu/people/Faculty/Sherman

http://www.rrfb.com

http://www.wormresearchcentre.co.uk

http://www.alternativeorganic.com

http://www.vermitech.com

www.jains.com/

www.techno-preneur.net/

www.hi-technaturalproducts.com

www.satavic.org/vermicomposting.htm -

www.indianindustry.com/fertilizers

www.dainet.org/livelihoods/vermicomposting

www.alanwood.net/pesticides/class

www.cabi.org/AbstractDatabases

www.ces.ncsu.edu/depts/hort/floriculture/software/pgr.html

anrcatalog.ucdavis.edu/PesticideUseSafety

www.wormdigest.org/content/

www.growingsolutions.com/home

www.dainet.org/livelihoods/vermicomposting

www.biocycle.net/BCArticles

www.manage.gov.in/agriclinics/getsst.asp

www.agmrc.org/agmrc/commodity/livestock/worms/

www.facebook.com/ads/

www.envfor.nic.in/legis/legis.html)

www.karmayog.org/library/libartdis.

www.e-pao.net/epSubPageSelector.asp

www.envindia.com/nandck/index.php/ecobazar_temp/comments

www.pscst.com/en/pub/index.htm -

INDEX

9 798223 885962